L8962

Construction technology

Volume 4

Second edition

R. Chudley M.C.I.O.B.

Chartered Builder

Illustrated by the author

Longman
Scientific &
Technical

51 6 14-99
6 90

Longman Scientific & Technical,
Longman Group UK Limited,
Longman House, Burnt Mill, Harlow,
Essex CM20 2JE, England
and Associated Companies throughout the world

First published 1977
Sixth impression 1985
Second edition 1987
Third impression 1989
Fourth impression 1991
Fifth impression 1992
Sixth impression 1993

British Library Cataloguing in Publication Data

Chudley, R.
 Construction technology.—2nd ed.
 Vol. 4
 1. Building
 I. Title
 690   TH145

ISBN 0-582-41396-6

General Editor
C. R. Bassett, B.Sc., F.C.I.O.B.
Formerly Principal Lecturer in the Department of Building an
Surveying, Guildford County College of Technology.

Set in IBM Journal 10 on 12 point

Produced by Longman Singapore Publishers (Pte) Ltd.
Printed in Singapore.

Contents

Introduction

This book completes the series of four volumes covering a typical four-year course in construction technology designed to supplement the student's lecture notes and project work. All the books have been written deliberately in a concise note form with ample illustrations firstly to fulfil their function as a supplement and secondly to try and keep the price down to a reasonable level. It has been assumed throughout the series that the bulk of the readers would be students following a definite pattern of study and I have therefore refrained from writing in depth on topics which are covered in allied subjects such as science, mathematics, materials and structures, structural theory and design, services, quantity surveying and administration.

No text-book or work of reference is ever complete in itself and I strongly recommend that students seek out all sources of reference on any particular topic of study to obtain the maximum amount of coverage which should lead to a better comprehension of the s bject. Building technology is not a pure academic subject, college lectures and text-books can only provide the necessary theoretical background to the building processes of design and works on site. Practical experience and/or observation of works in progress is therefore a vital ingredient of any comprehensive course of study in the subject of construction technology.

Acknowledge-ments

We are grateful to the following for permission to reproduce copyright material:

British Standards Institution for reference to British Standards, Codes of Practice; Building Research Establishment for extracts from Building Research Establishment Digests; Her Majesty's Stationery Office for extracts from Acts, Regulations and Statutory Instruments; the Electricity Council for extracts from their publication *Lighting for Building Sites*.

AUTHOR'S ACKNOWLEDGEMENTS

I should like to express my gratitude to the many students both past and present who have knowingly or unknowingly, by their comments, largely dictated the format and treatment of the contents of the four volumes, also to my colleagues at the Guildford County College of Technology for their valued comments, criticism and encouragement. Finally I wish to record my thanks to my daughter Susan and Mrs D. Boldero for their noble efforts with the typewriter in preparing the manuscripts, without whose help I should still be tapping away with two fingers to this day.

Roy Chudley
1977

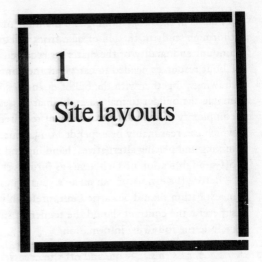

1
Site layouts

Part I
Site works

The construction of a building can be considered as the product being produced with a temporary factory, the building site being the 'factory' in which the building contractor will make the product. To enable this activity to take place the builder requires men, materials and plant, all of which have to be carefully controlled so that the men have the right machines in the most advantageous position, the materials stored so that they are readily available and not interfering with the general site circulation, and adequate storage space and site accommodation.

There is no standard size ratio between the free site space required to construct a building and the total size of the site on which the building is to be erected; therefore each site must be considered as a separate problem in terms of allocating space for men, materials and plant. To obtain maximum efficiency there is an optimum way of laying out the site and also a correct amount of expenditure to support the proposed site layout. Any planned layout should be reviewed periodically and adjusted to suit the changing needs of the site activities. If this aspect of building construction is carefully considered, planned and controlled it will be reflected in the progress and profitability of the contract.

Before any initial planning of the site layout can take place certain preliminary work must be carried out, preferably at the pre-tender stage. The decision to tender will usually be taken by the managing director or for small works by the senior estimator up to a contract value laid down by the managing director. With given designs and specifications the best opportunity for the contractor to prepare a competitive and economic tender is in the programming and planning of the construction activities. A thorough study of the bill of quantities will give an indication of the amount and quality of the materials required and also of the various labour resources needed to carry out the contract. A similar study of the drawings, together with the bill of quantities and the specification will enable the builder to make a preliminary assessment of the size and complexity of the contract, the plant required and the amount of money which can reasonably be expended on labour-saving items such as concrete mixing and placing alternatives, handling and transporting equipment and off-site fabrication of such items as formwork and reinforcement.

Before the estimator can make a start on calculating his unit rates a site investigation should be carried out, preferably by the site agent who will supervise the contract should the tender be successful. His report should include the following information:

1. *Access to site* — on- and off-site access, road and rail facilities, distances involved, rights of way restrictions, local authority or police restrictions and bridge weight or height limitations on approach routes.
2. *Services* — available power and water supplies together with rates of payment, nuisance or value of services already on site, diversions required and the time element involved in carrying out any necessary diversions together with cost implications.
3. *Layout* — general site conditions such as nature of soil, height of water table, flooding risks, tidal waters, neighbouring properties and any demolition problems.
4. *Labour* — travel distances, local or own labour resources to be used, availability of local labour and prevailing rates of pay, lodging and local catering facilities.
5. *Security* — local vandalism and pilfering record, security patrol facilities, need for night security, fencing and hoarding requirements.

With the knowledge and data gained from contract documents, site investigations and any information gained from the police and local authority sources the following pre-tender work can now be carried out:

1. *Pre-tender programme* — usually in a bar chart form showing the proposed time allowances for the major activities.
2. *Cost implications* — several programmes for comparison should be made to establish possible break-even points giving an indication of required overdraft, possible cash inflow and anticipated profit.
3. *Plant schedule* — can be prepared in the form of a bar chart showing requirements and utilisation which will help in deciding whether

there is a need for a site workshop and the necessary maintenance staff and equipment required on site. The problem of whether the plant should be purchased or hired, or if a balance of buying and hiring is most economical, will have to be considered at this stage.

4. *Materials schedule* — basic data can be obtained from the bill of quantities. The buyer's knowledge of the prevailing market conditions and future trends will enable usage and delivery periods and the amount of site space and/or accommodation required to be predicted.

5. *Labour summary* — basic data obtained from the bill of quantities, site investigation report and pre-tender bar chart programme to establish number and trades of staff required for future staff job allocation and also the amount and type of site accommodation required. Approximate labour requirements can be calculated as in the typical example below:

   Item — excavate by hand foundation trenches in ordinary ground commencing at oversite level and not exceeding 1.5 m deep.

   Assumed labour constant 2.5 hours per m^3.

   Total excavation quantity 525 m^3.

   Time allowed on pre-tender bar chart programme 1 month = 20 working days.

   Total man hours = 525 × 2.5 = 1 312.5

   Man hours per day = 1 312.5 ÷ 20 = 65.655

   Therefore assuming an 8-hour working day

   Number of men = 65.655 ÷ 8 = 8.2

   Say 9-man gang.

6. *Site organisation structure* — this is a 'family tree' chart showing the relationships and inter-relationships between the various members of the site team and is normally only required on large sites where the areas of responsibility and accountability must be clearly defined.

7. *Site layout* — site space allocation for materials storage, working areas, units of accommodation, plant positions and general circulation areas.

PLANNING SITE LAYOUTS

When planning site layouts the following must be taken into account:

1. Site activities.
2. Efficiency.
3. Movement.
4. Control.
5. Accommodation for staff and storage of materials.

Site activities: the time needed for carrying out the principal activities can be estimated from the data obtained previously for preparing the material and labour requirements. With repetitive activities estimates will be required to determine the most economical balance of units which will allow simultaneous construction processes; this in turn will help to establish staff numbers, work areas and material storage requirements. A similar argument can be presented for overlapping activities. If a particular process presents a choice in the way the result can be achieved the alternatives must be considered; for example, the rate of placing concrete will be determined by the output of the mixer and the speed of transporting the mix to the appropriate position. Alternatives which can be considered are:

1. More than one mixer.
2. Regulated supply of ready mixed concrete.
3. On large contracts, pumping the concrete to the placing position.

All alternative methods for any activity will give different requirements for staff numbers, material storage, access facilities and possibly plant types and numbers.

Efficiency: to achieve maximum efficiency the site layout must aim at maintaining the desired output of the planned activities throughout the working day and this will depend largely upon the following factors:

1. Avoidance, as far as practicable, of double handling materials.
2. Proper store-keeping arrangements to ensure that the materials are of the correct type, in the correct quantity and are available when required.
3. Walking distances are kept to a minimum to reduce the non-productive time spent in covering the distances between working, rest and storage areas without interrupting the general circulation pattern.
4. Avoidance of loss by the elements by providing adequate protection for unfixed materials on site, thereby preventing time loss and cost of replacing damaged materials.
5. Avoidance of loss by theft and vandalism by providing security arrangements in keeping with the value of the materials being protected and by making the task difficult for the would-be thief or vandal by having adequate hoardings and fences. Also to be avoided is the loss of materials due to pilfering by site staff who may consider this to be a perquisite of the industry. Such losses can be reduced by having an adequate system of stores' requisition and material checking procedures.
6. Minimising on-site traffic congestion by planning delivery arrivals,

having adequate parking facilities for site staff cars and mobile machinery when not in use, and by having sufficient turning circle room for the types of delivery vehicles likely to enter the site.

Movement: apart from the circulation problems mentioned above the biggest problem is one of access. Vehicles delivering materials to the site should be able to do so without difficulty or delay. Many of the contractors' vehicles will be lightweight and will therefore present few or no problems, but the weight and length of suppliers' vehicles should be taken into account. For example, a fully laden ready-mix concrete lorry can weigh 20 tonnes and lorries used for delivering structural steel can be 18.000 metres long, weighing up to 40 tonnes and requiring a large turning circle. If it is anticipated that heavy vehicles will be operating on site it will be necessary to consider the road surface required. If the roads and paved areas are part of the contract and will have adequate strength for the weight of the anticipated vehicles it may be advantageous to lay the roads at a very early stage in the contract, but if the specification for the roads is for light traffic it would be advisable to lay only the base hoggin or hardcore layer at the initial stages because of the risk of damage to the completed roads by the heavy vehicles. As an alternative it may be considered a better policy to provide only temporary roadways composed of railway sleepers, metal tracks or mats until a later stage in the contract, especially if such roads will only be required for a short period.

Control: this is mainly concerned with the overall supervision of the contract, including men, materials and the movement of both around the site. This control should form the hub of the activities which logically develops into areas or zones of control radiating from this hub or centre. Which zone is selected for storage, accommodation or specific activities is a matter of conjecture and the conditions prevailing on a particular site but as a rule the final layout will be one of compromise with storage and accommodation areas generally receiving priority.

Accommodation: as previously stated this must be considered for each individual site but certain factors will be common to all sites. Accommodation for staff is covered by the Construction (Health and Welfare) Regulations 1966, the main contents of which should have been studied in the second year (see Chapter 1, Volume 2). This document sets out the minimum amount and type of accommodation which must legally be provided for the number of persons employed on the site and the anticipated duration of the contract. Apart from these minimum requirements the main areas of concern will be sizing, equipping and siting the various units of accommodation.

Mess huts: covered by Regulation No. 11 and are for the purposes of preparing, heating and consuming food which may require the following services: drainage, light, power, hot and cold water supply. To provide a reasonable degree of comfort a floor area of 2.0 to 2.5 m^2 per person should be allowed. This will provide sufficient circulation space and room for tables, seating and space for the storage of any utensils. Consideration can also be given to introducing a system of staggered meal-breaks thus reducing space requirements. On large sites where full canteen facilities are being provided it may be prudent to place this in the hands of a catering firm. Mess huts should be sited so that they do not interfere with the development of the site but in such a position that travel time is kept to a minimum. On sites which by their very nature are large, it is worth while considering a system whereby tea-breaks can be taken in the vicinity of the work areas. Siting mess huts next to the main site circulation and access roads is not of major importance.

Drying rooms: used for the purposes of depositing and drying wet clothes are covered by Regulation No. 11. Drying rooms generally require a lighting and power supply with lockers or racks for deposited clothes. A floor area of 0.6 m^2 per person should provide sufficient space for equipment and circulation. Drying rooms should be sited near or adjacent to the mess room.

Toilets: contractors are required to provide at least the necessary minimum washing and sanitary facilities as set out in Regulations Nos. 12, 13 and 14. All these facilities will require light, water and drainage services. If it is not possible or practicable to make a permanent or temporary connection to a drainage system the use of chemical methods of disposal should be considered. Sizing of toilet units is governed by the facilities being provided and if female staff are employed on site separate toilet facilities must be provided. Toilets should be located in a position which is convenient to both offices and mess rooms, which may mean providing more than one location on large sites.

First-aid rooms: only required on large sites where the number of persons employed exceeds 250, but any contractor who has more than 40 persons in his employment on that site must provide this facility in accordance with the requirements of Regulation No. 9. The first-aid room should be sited in a position which is conveniently accessible from the working areas and must be of such a size as to allow for the necessary equipment and adequate circulation which would indicate a minimum floor area of 6 m^2.

Before the proposed site layout is planned and drawn the contracts manager and the proposed site agent should visit the site to familiarise themselves with the prevailing conditions. During this visit the position and condition of any existing roads should be noted and the siting of any temporary roads considered necessary should be planned. Information regarding the soil conditions, height of water table, and local weather patterns should be obtained by observation, site investigation, soil investigation, local knowledge or from the local authority. The amount of money which can be expended on this exercise will depend upon the size of the proposed contract and possibly upon how competitive the tenders are likely to be for the contract under consideration.

Figure I.1 shows a typical small-scale general arrangement drawing and needs to be read in conjunction with Fig. I.2 which shows the proposed site layout. The following data has been collected from a study of the contract documents and by carrying out a site investigation:

1. Site is in a typical urban district within easy reach of the contractor's main yard and therefore will present no transport or labour availability problems.
2. Subsoil is a firm sandy clay with a water table at a depth which should give no constructional problems.
3. Possession of site is to be at the end of April and the contract period is 18 months. The work can be programmed to enable the foundation and substructure work to be completed before adverse winter weather conditions prevail.
4. Development consists of a single five-storey office block with an *in situ* reinforced concrete structural frame, *in situ* reinforced concrete floors and roof, precast concrete stairs and infill brick panels to the structural frame with large hardwood timber frames fixed into openings formed by the bricklayers. Reduced level dig is not excessive but the top soil is to be retained for landscaping upon completion of the building contract by a separate contractor; the paved area in front of the office block, however, forms part of the main contract. The existing oak trees in the north-east corner of the site are to be retained and are to be protected during the contract period.
5. Estimated maximum number of staff on site at any one time is 40 in the ratio of 1 supervisory staff to 10 operatives plus a resident Clerk of Works.
6. Main site requirements are as follows:
   1 office for 3 supervisory staff.
   1 office for resident Clerk of Works.
   1 office for timekeeper and materials checker.
   1 hutment as lock-up store.

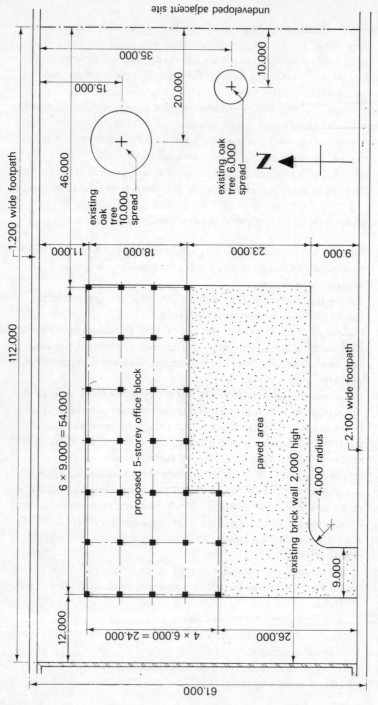

undeveloped adjacent site

1.200 wide footpath

112.000

35.000

15.000

20.000

46.000

existing oak tree 10.000 spread

existing oak tree 6.000 spread

10.000

N

11.000

18.000

23.000

9.000

6 × 9.000 = 54.000

proposed 5-storey office block

2.100 wide footpath

paved area

existing brick wall 2.000 high

4.000 radius

12.000

4 × 6.000 = 24.000

26.000

9.000

61.000

Fig I.1 Site layout example — general arrangement

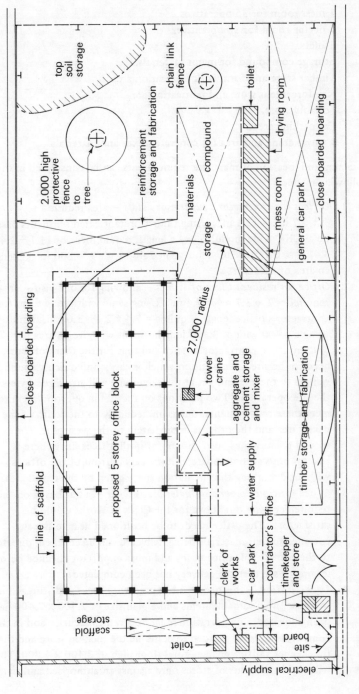

Fig 1.2 Site layout example — proposed layout of accommodation and storage

9

1 mess room for 36 operatives.
1 drying room for 36 operatives.
Toilets.
Storage compound for major materials.
Timber store and formwork fabrication area.
Reinforcement store and fabrication area.
Scaffold store.
Car parking areas.
1 tower crane and area for concrete mixer and materials.

Sizing and location of main site requirements can be considered in the following manner:

1. *Offices for contractor's supervisory staff* — area required = $3 \times 3.7 \text{ m}^2 = 11.1 \text{ m}^2$. Using timber prefabricated hutments based on a 2.400 wide module gives a length requirement of $11.1 \div 2.4 = 4.625$ m, therefore use a hutment 2.400 wide $\times$ 4.800 long giving an area of 11.52 m^2.

2. *Office for resident Clerk of Works* — allowing for one visitor area required = $2 \times 3.7 \text{ m}^2 = 7.4 \text{ m}^2$. Using same width module as for contractor's office length required = $7.4 \div 2.4 = 3.08$ m, therefore using a 2.400 wide $\times$ 3.300 long hutment will give an area of 7.92 m^2. The contractor's office and that for the Clerk of Works needs to be sited in a position which is easily and quickly found by visitors to the site and yet at the same time will give a good view of the site operations. Two positions on the site in question seem to meet these requirements: one is immediately to the south of the paved area and the other is immediately to the west of it. The second position has been chosen for both offices since there is also room to accommodate visitors' cars in front of the offices without disturbing the circulation space given by the paved area.

3. *Office for timekeeper and materials checker* — a hut based on the requirements set out above for the Clerk of Works would be satisfactory. The office needs to be positioned near to the site entrance so that materials being delivered can be checked, directed to the correct unloading point and most important checked before leaving to see that the delivery has been completed.

4. *Lock-up store* — this needs to be fitted with racks and storage bins to house small valuable items and a plan size of 2.400 x 2.400 has been allocated. Consideration must be given to security and in this context it has been decided to combine the lock-up store and the timekeeper's office giving a total floor plan of 2.400 x 4.800. This will enable the issue of stores only against an authorised and signed

requisition to be carefully controlled, the timekeeper fulfilling the function of storekeeper.

5. *Mess room* — area required = 36 × 2.5 m^2 = 90 m^2, using a width module of 4.800, length required = 90 ÷ 4.8 = 18.75 m; therefore using a length of 7.5 modules of 2.400 actual length = 18.000 giving an area of 4.8 × 18.0 = 86.4 m^2 which is considered satisfactory. The mess room needs to be sited in a fairly central position to all the areas of activity and the east end of the paved area has been selected.

6. *Drying room* — area required = 36 × 0.6 m^2 = 21.6 m^2, using a width module of 4.800, length required = 21.6 ÷ 4.8 = 4.5 m; therefore using a length of 2 modules of 2.400 actual length = 4.800, giving an area of 4.8 × 4.8 = 23.04 m^2. The drying room needs to be in close proximity to the mess room and has therefore been placed at the east end of the mess room. Consideration could be given to combining the mess room and drying room into one unit.

7. *Toilets* — a decision was made during the site investigation to hire and use self-contained chemical toilets to eliminate the need for temporary drains thus giving complete freedom in the programming of drainage works. Two such units are considered to be adequate, one to be sited near to the mess room and the other to be sited near to the office complex. The minimum number of sanitary conveniences is laid down in Regulation No. 13 in the Construction (Health and Welfare) Regulations 1966. For the mess toilet unit catering for 36 operatives, two conveniences are required as a minimum but a three-convenience toilet unit will be used having a plan size of 2.400 × 3.600. Similarly although only one convenience is required for the office toilet unit a two-convenience unit will be used with a plan size of 2.400 × 2.400.

8. *Materials storage compound* — area to be defined by a temporary timber fence 1.800 high and sited at the east end of the paved area giving good access for deliveries and within reach of the crane. Plan size to be allocated 12.000 wide × 30.000 long.

9. *Timber storage* — timber is to be stored in top-covered but open-sided racks made from framed standard scaffold tubulars. Maximum length of timber to be ordered is unlikely to exceed 6.000 in length; therefore, allowing for removal, cutting and fabricating into formwork units, a total plan size of 6.000 wide × 36.000 long has been allocated. This area has been sited to the south of the paved area, giving good access for delivery and within the reach of the crane.

10. *Reinforcement storage* — the bars are to be delivered cut to length,

bent and labelled and will be stored in racks as described above for timber storage. Maximum bar length to be ordered assumed not to exceed 12.000; therefore a storage and fabrication plan size of 6.000 wide x 30.000 long has been allocated. This area has been sited to the north of the storage compound, giving reasonable delivery access and within reach of the crane.

11. *Scaffold storage* — tube lengths to be stored in racks as described for timber storage with bins provided for the various types of couplers. Assuming a maximum tube length of 6.000 a plan size of 3.000 wide x 12.000 long. This storage area has been positioned alongside the west face of the proposed structure, giving reasonable delivery access and within reach of the crane if needed. The scaffold to be erected will be of an independent type around the entire perimeter positioned 200 mm clear of the building face and of five-board width, giving a total minimum width of 200+ (5 x 225) = 1.325, say 1.400 total width.

12. *Tower crane* — to be sited on the paved area in front of the proposed building alongside the mixer and aggregate storage position. A crane with a jib length of 27.000, having a lifting capacity of 1.25 tonnes at its extreme position, has been chosen so that the crane's maximum radius will cover all the storage areas thus making maximum utilisation of the crane possible.

13. *Car parking* — assume 20 car parking spaces are required for opera-tives needing a space per car of 2.300 wide x 5.500 long, giving a total length of 2.3 x 20 = 46.000 and, allowing 6.000 clearance for manoeuvring, a width of 5.5 + 6.0 = 11.500 will be required. This area can be provided to the south of the mess room and drying room complex. Staff car parking space can be sited in front of the office hutments giving space for the parking of seven cars which will require a total width of 7 x 2.3 = 16.100.

14. *Fencing* — the north and south sides of the site both face onto public footpaths and highways. Therefore a close-boarded hoarding in accordance with the licence issued by the local authority will be provided. A lockable double-gate is to be included in the south side hoarding to give access to the site. The east side of the site faces an undeveloped site and the contract calls for a 2.000 high concrete post and chain-link fence to this boundary. This fence will be erected at an early stage in the contract to act as a security fence during the construction period as well as providing the permanent fencing. The west side of the site has a 2.000 high brick wall which is in a good structural condition and therefore no action is needed on this boundary.

15. *Services* — it has been decided that temporary connections to the foul drains will not be made thus giving complete freedom in planning the drain-laying activities. The permanent water supply to the proposed office block is to be laid at an early stage and this run is to be tapped to provide the supplies required to the mixer position and the office complex. A temporary connection is to be made to supply the water service to the mess room complex since a temporary supply from the permanent service would mean running the temporary supply for an unacceptable distance. An electrical supply is to be taken onto site with a supply incoming unit housed in the timekeeper's office along with the main distribution unit. The subject of electrical supplies to building sites is dealt with in Chapter 2. Telephones will be required to the contractor's and Clerk of Works' offices. It has been decided that a gas supply is not required.

16. *Site identification* — a V-shaped board bearing the contractor's name and company symbol is to be erected in the south-west corner of the site in such a manner that it can be clearly seen above the hoarding by traffic travelling in both directions enabling the site to be clearly identified. The board will also advertise the company's name and possibly provide some revenue by including on it the names of participating sub-contractors. As a further public relations exercise it might be worth while considering the possibility of including public viewing panels in the hoarding on the north and south sides of the site.

The extent to which the above exercise in planning a site layout would be carried out in practice will depend upon a number of factors such as the time and money which can reasonably be expended and the benefits which could accrue in terms of maximum efficiency compared with the amount of the capital outlay. The need for careful site layout and site organisation planning becomes more relevant as the size and complexity of the operation increases. This is particularly true for contracts where spare site space is very limited.

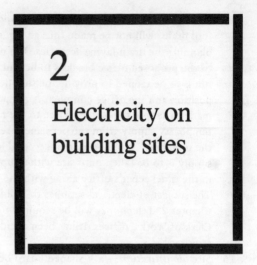

2
Electricity on building sites

A supply of electricity is usually required on construction sites to provide lighting to the various units of accommodation and may also be needed to provide the power to drive small and large items of plant. Two sources of electrical supply to the site are possible, namely:

1. Portable self-powered generators.
2. Metered supply from the local area electricity board.

Since a supply of electricity is invariably required in the final structure the second source is usually adopted because it is generally possible to tap off the permanent supply cable to the proposed development for construction operations, thus saving the cost of laying a temporary supply cable to the site.

To obtain a metered temporary supply of electricity a contract must be signed between the contractor and the local area electricity board who will require the following information:

1. Address of site.
2. Site location plan.
3. Maximum anticipated load demand in kW for the construction period. A reasonable method of estimating this demand is to allow for a loading of 10 W/m² for the total floor area of the finished structure and to add for any high load equipment such as cranes, pumps and drying out heaters which are to be used.

4. Final load demand of the completed building to ensure that the correct rating of cable is laid for the permanent supply.
5. Date on which temporary supply will be required.
6. Name, address and telephone number of the building owner and the contractor.

To ensure that the supply and installation is available when required by the builder it is essential that an application for a temporary supply of electricity is made at the earliest possible date.

On any construction site it is possible that there may be existing electricity cables which can be advantageous or constitute a hazard or nuisance. Overhead cables will be visible whereas the routes and depths of underground cables can only be ascertained from the records and maps kept by the local area supply board. Overhead cable voltages should be checked with the local area supply board since these cables are usually uninsulated and are therefore classed as a hazard due mainly to their ability to arc over a distance of several metres. High voltage cables of over 11 kV rating will need special care and any of the following actions could be taken to reduce or eliminate the danger:

1. Apply to the local area supply board to have the cables re-routed at a safe distance or height.
2. Apply to have the cable taken out of service.
3. Erect warning barriers to keep men and machines at a safe distance. These barriers must be clearly identified as to their intention and they may be required to indicate the safe distance in both the horizontal and vertical directions. The local area supply board will advise on suitable safe distances according to the type of cable and the load it is transmitting.

The position and depth of underground cables given by the local area board must be treated as being only approximate since their records show only the data regarding the condition as laid and since when changes in site levels may have taken place. When excavating in the locality of an underground cable extreme caution must be taken, which may even involve careful hand excavation to expose the cable. Exposed cables should be adequately supported and suitable barriers with warning notices should be erected. Any damage, however minor, must be reported to the local area supply board who will take the necessary remedial action. It is worth noting that if a contractor damages an underground electric cable, which he knew to be present, and possibly causing a loss of supply to surrounding properties he can be sued for negligence and trespass to goods, the remedy for both torts is an action for damages.

SUPPLY AND INSTALLATION

In Great Britain electrical installations on construction sites are subject to the requirements of the Electricity (Factories Act) Special Regulations 1908 and 1944. Requirements concerning precautions to be taken against contact with overhead lines and underground cables encountered on site are contained in the Construction (General Provisions) Regulations 1961 Part XI Regulation 44. The installation should follow the rules given in the Regulations for Electrical Equipment of Buildings issued by the Institute of Electrical Engineers and in particular Part H which deals with temporary installations and installations on construction sites. The supply distribution units used in the installation should comply with the recommendations of BS 4363 which covers the equipment suitable for the control and distribution of electricity from a three-phase four-wire a.c. system up to a voltage of 415 V with a maximum capacity of 300 A per phase.

The appliances and wiring used in temporary installations on construction sites may be subject to extreme abuse and adverse conditions; therefore correct circuit protection, earthing and frequent inspection are most important and this work, including the initial installation, should be entrusted to a qualified electrician or to a specialist electrical contractor.

Electrical distribution cables contain three line wires and one neutral which can give either a 415 V three-phase supply or a 240 V single-phase supply. Records of accidents involving electricity show that the highest risk is encountered when electrical power is used in wet or damp conditions, which are often present on construction sites. It is therefore generally recommended that wherever possible the distribution voltage on building sites should be 110 V. This is a compromise between safety and efficiency but it cannot be over stressed that a supply of this pressure can still be dangerous and lethal.

The recommended voltages for use on construction sites are as follows:

Mains voltage

415 V three-phase	supply to transformer unit, heavy plant such as cranes and movable plant fed via a trailing cable
	hoists and plant powered by electric motors in excess of a 2 kW rating
240 V single-phase	supply to transformer unit
	supply to distribution unit
	installations in site accommodation buildings
	fixed floodlighting
	small static machines

Reduced voltage

110 V three-phase   portable and hand held tools
110 V single-phase   portable and hand held tools
                            small items of plant
                            site floodlighting other than fixed floodlighting
                            portable hand-lamps
50 V single-phase ⎫   as listed for 110 V single-phase but being used in
25 V single-phase ⎭   damp situations

It is worth considering the use of 50 or 25 V battery supplied hand-lamps if damp situations are present on site. All supply cables must be earthed and in particular 110 V supplies should be centre point earthed so that the nominal voltage to earth is not more than 65 V on a three-phase circuit and not more than 55 V on a single-phase circuit.

Protection to a circuit can be given by using bridge fuses, cartridge fuses and circuit breakers. Adequate protection should be given to all main and sub-circuits against any short-circuit current, overload current and earth faults.

Protection through earthing may be attained in two distinct ways:

1. Provision of a path of low impedence to ensure over-current device will operate in a short space of time.
2. Insertion in the supply of a circuit breaker with an operating coil which trips the breaker when the current due to earth leakage exceeds a predetermined value.

BS 4363 recommends that plug and socket outlets are identified by a colour coding as an additional safety precaution to prevent incorrect connections being made, the recommended colours are:

     25 V − violet
     50 V − white
  110 V − yellow
  240 V − blue
  415 V − red

The equipment which can be used to distribute an electrical supply around a construction site is as follows:

1. *Supply incoming unit* (SIU) − supply, control and distribution of mains supply on site − accommodates supply board's equipment and has one outgoing circuit.
2. *Main distribution unit* (MDU) − control and distribution of mains supply for circuits of 415 V three-phase and 240 V single-phase.
3. *Supply incoming and distribution unit* (SIDU) − a combined SIU

and MDU for use on sites where it is possible to locate these units together.

4. *Transformer unit* — transforms and distributes electricity at a reduced voltage and can be for single-phase, three-phase or both phases and is abbreviated TU/1, TU/3 or TU/1/3 accordingly.

5. *Outlet unit* (OU/1 or OU/3) — connection, protection and distribution of final sub-circuits at a voltage lower than the incoming supply.

6. *Extension outlet unit* (EOU/1 or EOU/3) — similar to outlet unit except outlets are not protected.

7. *Earth monitor unit* (EMU) — flexible cables supplying power at mains voltage from the MDU to movable plant incorporate a separate pilot conductor in addition to the main earth continuity conductor. A very low-voltage current passes along these conductors between the portable plant and the fixed EMU. A failure of the earth continuity conductor will interrupt the current flow which will be detected by the EMU and this device will automatically isolate the main circuit.

The cubicles or units must be of robust construction, strong, durable, rain resistant and rigid to resist any damage which could be caused by transportation, site handling or impact shocks likely to be encountered on a construction site. All access doors or panels must have adequate weather seals. Figure I.3 shows a typical supply and distribution system for a construction site.

The routing of the supply and distribution cables around the construction site should be carefully planned. Cables should not be allowed to trail along the ground unless suitably encased in a tube or conduit, and even this method should only be used for short periods of time. Overhead cables should be supported by hangers attached to a straining wire and suitably marked with 'flags' or similar visual warning. Recommended minimum height clearances for overhead cables are:

1. 5.200 in positions inaccessible to vehicles.
2. 5.800 where cable crosses an access road or any part of the site accessible to vehicles.

Cables which are likely to be in position for a long time, such as the supply to a crane, should preferably be sited underground at a minimum depth of 500 mm and protected by tiles or alternatively housed in clayware or similar pipes.

In the interest of safety and to enable first-aid treatment to be given in cases of accident, all contractors using a supply of electricity on a construction site for any purpose must display in a prominent position the

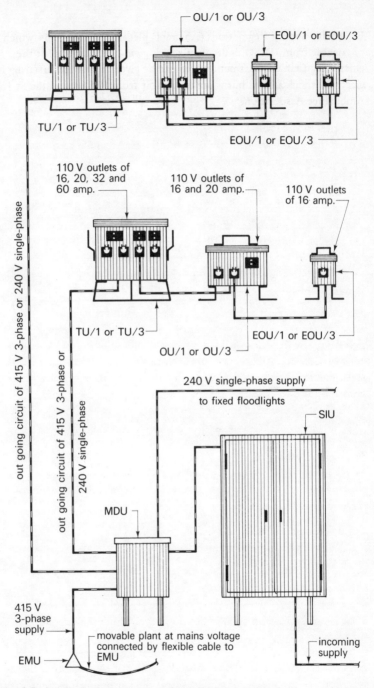

Fig I.3 Typical distribution sequence of site electricity

Electricity Special Regulations (Abstract) placard number F954 which is obtainable from HMSO. Suitable placards giving instructions for the emergency first-aid treatment which can be given to persons suffering from electrical shock and/or burns are obtainable from RoSPA and the St John Ambulance Association.

3
Lighting building sites

Inadequate light accounts for more than 50% of the loss of production on construction sites between the months of November and February. Inadequate lighting also increases the risks of accidents and lowers the security of the site. The initial costs of installing a system of artificial lighting for both internal and external activities can usually be offset by higher output, better quality work, a more secure site and apportioning the costs over a number of contracts on a use and re-use basis.

The reasons for installing a system of artificial lighting on a construction site can be listed as follows:

1. Inclement weather, particularly in winter when a reduction of natural daylight is such that the carrying out of work becomes impracticable.
2. Without adequate light all activities on construction sites carry an increased risk of accident and injury.
3. By enabling work to proceed losses in productivity can be reduced.
4. Reduce the wastage of labour and materials which often results from working in poor light.
5. Avoid short-time working due to the inability to see clearly enough for accurate and safe working.
6. Improve the general security of the site.

The following benefits may be obtained by installing and using a system of artificial lighting on a construction site:

21

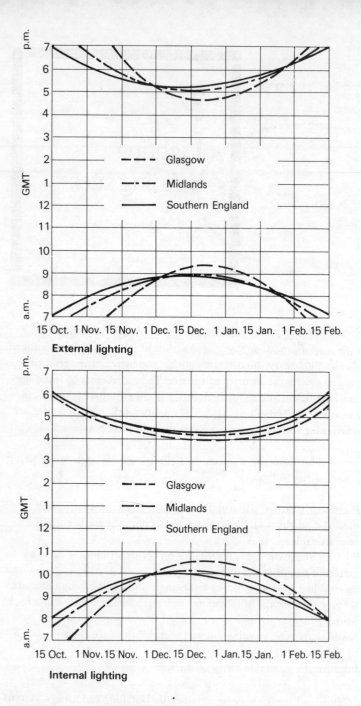

External lighting

Internal lighting

Fig I.4 Approximate times for site lighting

1. Site activities will be independent of the availability of natural day-light and therefore the activities can be arranged to suit the needs of the contract, the availability of materials and the personnel involved.
2. Overtime and extra shifts can be worked to overcome delays that might occur from any cause.
3. Deliveries and collection of materials or plant can be made outside normal site working hours thus helping to avoid delays and/or congestion.
4. Reduction in the amount of spoilt material and the consequent rectification caused by working under inadequate light.
5. An effective deterrent to the would-be trespasser or pilferer.
6. Improved labour relationships by ensuring regular working hours and thus regular earnings.

Planning the lighting requirements depends on site layout, size of site, shape of site, geographical location, availability of an electrical supply and the planned activities for the winter period. Figure I.4 shows two charts covering various regions of the country giving an indication of the periods when external and internal artificial lighting may be required on a construction site under normal conditions. Any form of temporary artificial site lighting should be easy to install and modify as needs change and be easy to remove whilst works are still in progress.

The supply and distribution of an electrical service to a construction site has already been covered in the previous chapter and it is therefore only necessary to stress again the need for a safe, reliable installation preferably designed and installed by a specialist contractor.

ILLUMINATION

Illumination is measured in lux (lx) which is one lumen of light falling on 1 m^2 of surface and this can be measured with a small portable lightmeter which consists of a light-sensitive cell generating a small current proportional to the light falling on it. The level of illumination at which an operative can work in safety and carry out his tasks to an acceptable standard, both in terms of speed and quality, is quite low since the human eye is very adaptable and efficient. Although the amount of illumination required to enable a particular activity to be carried out is a subjective measure depending largely upon the task, age and state of health of the operative concerned, the following average service values of illumination are generally recommended:

Activities	Illumination (lx)
External lighting	
Materials handling	10
Open circulation areas	10
Internal lighting	
Circulation	5
Working areas	15
Reinforcing and concreting	50
Joinery, bricklaying and plastering	100
Painting and decorating	200
Fine craft work	200
Site offices	200
Drawing board positions	300

The above service values of illumination do not allow for deterioration, dirt, bad conditions or shadow effects. Therefore in calculating the illumination required for any particular situation a target value of twice the service value should be used.

When deciding on the type of installation to be used two factors need to be considered:

1. Type of lamp to be used.
2. Nature and type of area under consideration.

The properties of the various types of lamps available should be examined to establish the most appropriate for any particular site requirement.

LAMPS

Tungsten filament lamp — ideal for short periods such as a total of 200 hours during the winter period; main recommended uses are for general interior lighting and low-level external movement. They are cheap to buy but are relatively expensive to run.

Tungsten halogen lamp — compact fitting with high light output and is suitable for all general area floodlighting. They are easy to mount and have a more effective focused beam than the filament lamp. These lamps generally have a life of twice that of filament lamps and quartz lamps have a higher degree of resistance to thermal shock than glass filament lamps. They are dearer than filament lamps and are still relatively expensive to run but should be considered if the running time is in the region of 1 500 hours annually.

Mercury tungsten lamps — compact, efficient with a good lamp life and they do not need the expensive starting gear of the vapour discharge lamps. They can be used for internal and external area lighting where lamps are not mounted above 9.000 high. These are a high-cost lamp but are cheap to run.

Mercury discharge lamps — high efficiency lamps with a long life and can be used for area lighting where lamps are mounted above 9.000 high. Costs for lamps and control gear are high but the running costs are low.

Tubular fluorescent lamps — uniformly bright in all directions, used when a great concentration of light is not required, efficient with a range of colour values. These lamps have a long life and are cheap to run.

High-pressure sodium discharge lamps — compact, efficient with a long life and for the best coverage without glare they should be mounted above 13.500 high. Cost for lamp and control gear is high but running costs are low which makes them suitable for area lighting.

Apart from the cost of the lamps and the running charges, consideration must be given to the cost of cables, controlling equipment, mounting poles, masts or towers. A single high tower may well give an overall saving against using a number of individual poles or masts in spite of the high initial cost for the tower. Consideration can also be given to using the scaffold, incomplete structure or the mast of a tower crane for lamp-mounting purposes.

SITE LIGHTING INSTALLATIONS
When deciding upon the type and installation layout for construction site lighting, consideration must be given to the nature of area and work to be lit, and also to the type or types of lamp to be used. These aspects can be considered under the following headings:

1. External and large circulation areas.
2. Beam floodlighting.
3. Walkway lighting.
4. Local lighting.

External and large circulation areas
These areas may be illuminated by using mounted lamps situated around the perimeter of the site, in the corners

of the site or, alternatively, overhead illumination using dispersive fittings
can be used. The main objectives of area lighting are to enable man and
machinery to move around the site in safety and to give greater security
to the site. Areas of local danger such as excavations and obstructions
should however be marked separately with red warning lights or amber
flashing lamps. Tungsten filament, mercury vapour or tungsten halogen
lamps can be used and these should be mounted on poles, masts or towers
according to the lamp type and wattage. Typical mounting heights for
various lamps and wattages are:

Lamp type	Watts	Minimum height (m)
Tungsten filament	200	4.500
	300	6.000
	750	9.000
Mercury vapour	400	9.000
	1 000	15.000
	2 000	18.000
Tungsten halogen	500	7.500
	1 000	9.000
	2 000	15.000

Large areas are generally illuminated by using large high-mounted
lamps whereas small areas and narrow sites use a greater number of smaller
fittings. By mounting the lamps as high as practicable above the working
level, glare is reduced and by lighting the site from at least two directions
the formation of dense shadows is also reduced. The spacing of the
lamps is also important if under-lit and over-lit areas are to be avoided.
Figures I.5 and 6 show typical lamps and the recommended spacing
ratios.

Dispersive lighting is similar to an ordinary internal overhead lighting
system and is suitable for both exterior and interior area lighting where
overhead suspension is possible. Ordinary industrial fittings should not be
used because of the adverse conditions which normally prevail on construction sites. The fittings selected should therefore be protected against rust,
corrosion and water penetration. To obtain a reasonable spread of light
the lamps should be suspended evenly over the area to be illuminated as
shown diagrammatically in Fig. I.7. Tungsten filament, mercury vapour
and fluorescent trough fittings are suitable and should be suspended at a
minimum height according to their type and wattage. Typical suspension
heights are:

Lamp type	Watts	Minimum height (m)
Tungsten filament	200	2.500
	300	3.000
	750	6.000
Mercury vapour	250	6.000
	400	7.500
	700	9.000
Fluorescent trough	40 to 125	2.500

Most manufacturers provide guidance as to the choice of lamps or combination of lamps but a simple method of calculating lamp requirements is as follows:

1. Decide upon the service illumination required and double this figure to obtain target value.

2. Calculate total lumens required $= \dfrac{\text{area } (m^2) \times \text{target value (lx)}}{0.23}$

3. Choose lamp type

   Number of lamps required $= \dfrac{\text{total lumens required}}{\text{lumen output of chosen lamp}}$

4. Repeat stage (3) for different lamp types to obtain most practicable and economic arrangement.

5. Consider possible arrangements remembering that:
   (a) Larger lamps give more lumens per watt and are generally more economic to run.
   (b) Fewer supports simplify wiring and aid overall economy.
   (c) Corner siting arrangements are possible.
   (d) Clusters of lamps are possible.

The calculations when using dispersive lighting are similar to those given above for mounted area lighting except for the formula in stage (2) which has a utilisation factor of 0.27 instead of 0.23.

Beam floodlighting

Tungsten filament or mercury vapour lamps can be used but the use of this technique is limited in application on construction sites to supplementing other forms of lighting. Beam floodlights are used to illuminate areas from a great distance. The beam of light is intense producing high glare and should therefore be installed to point downwards towards the working areas. Generally the lamps are selected direct from the manufacturers' catalogue without calculations.

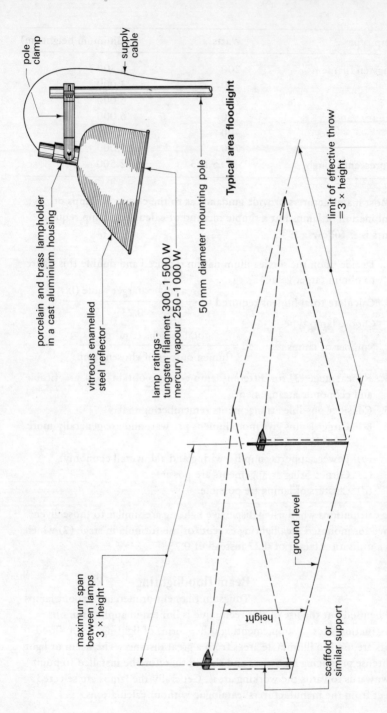

pole clamp

supply cable

Typical area floodlight

porcelain and brass lampholder in a cast aluminium housing

vitreous enamelled steel reflector

lamp ratings – tungsten filament 300 – 1 500 W mercury vapour 250 – 1 000 W

50 mm diameter mounting pole

limit of effective throw = 3 × height

ground level

maximum span between lamps 3 × height

height

scaffold or similar support

Fig I.5 Area lighting — lamps and spacing ratios 1

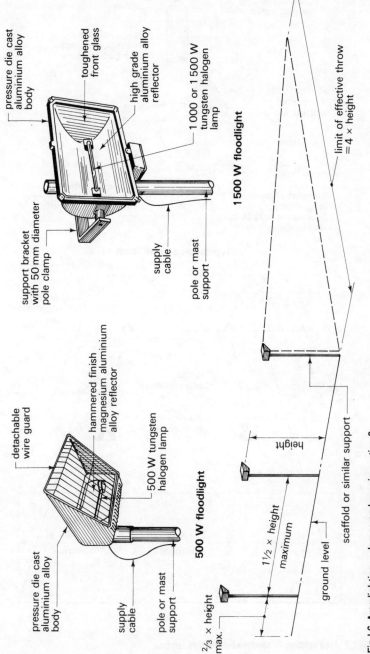

1 500 W floodlight

pressure die cast aluminium alloy body

toughened front glass

high grade aluminium alloy reflector

1 000 or 1 500 W tungsten halogen lamp

support bracket with 50 mm diameter pole clamp

supply cable

pole or mast support

500 W floodlight

detachable wire guard

hammered finish magnesium aluminium alloy reflector

500 W tungsten halogen lamp

pressure die cast aluminium alloy body

supply cable

pole or mast support

²⁄₃ × height max.

1½ × height maximum

height

ground level

scaffold or similar support

limit of effective throw ≡ 4 × height

Fig I.6 Area lighting — lamps and spacing ratios 2

29

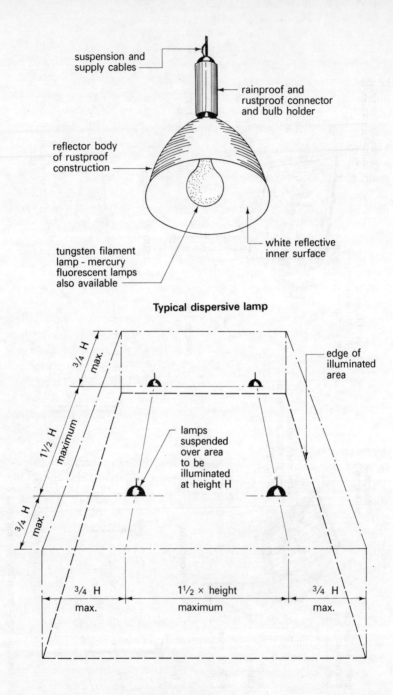

suspension and supply cables

rainproof and rustproof connector and bulb holder

reflector body of rustproof construction

white reflective inner surface

tungsten filament lamp - mercury fluorescent lamps also available

Typical dispersive lamp

edge of illuminated area

³/₄ H max.

1½ H maximum

lamps suspended over area to be illuminated at height H

³/₄ H max.

³/₄ H max.

1½ × height maximum

³/₄ H max.

Fig I.7 Area lighting — overhead dispersive lamps

Walkway lighting

Tungsten filament and fluorescent lamps can be used to illuminate access routes such as stairs, corridors and scaffolds. Bulkhead fittings which can be safely installed with adequate protection to the wiring can be run off a mains voltage of 240 V single-phase but if they are in a position where they can be handled a reduced voltage of 110 V single-phase should be used. Festoon lighting in which the ready-wired lampholders are moulded to the cable itself can also be used. A standard festoon cable would be 100 m long with rain-proof lampholders and protective shades or guards at 3.000 or 5.000 centres using 40 W or 60 W tungsten filament bulbs for the respective centres. See Fig. I.8. For lighting to scaffolds of 4 or 5 board width 60 W lamps placed at not more than 6.000 centres and preferably at least 2.400 to 3.000 above the working platform either to the wall side or centrally over the scaffold.

Local lighting

Clusters of pressed glass reflector flood-lamps, tungsten filament lamps, festoons and adjustable fluorescent can be used to increase the surface illumination at local points, particularly where finishing trades are involved. These fittings must be portable so that shadow casting can be reduced or eliminated from the working plane; therefore it is imperative that these lights are operated off a reduced voltage of 110 V single-phase. Fluorescent tubes do not usually work at a reduced voltage so special fittings working off a 110 V single-phase supply which internally increase the voltage are used. Typical examples of suitable lamps are shown in Fig. I.9.

As an alternative to a system of static site lighting connected to the site mains electrical supply mobile lighting sets are available. These consist of a diesel engine driven generator and a telescopic tower with a cluster of tungsten iodine lamps. These are generally cheaper to run than lamps operating off a mains supply. Small two-stroke generator sets with a single lamp attachment suitable for small isolated positions are also available.

Another system which can be used for local lighting is flame lamps which normally use propane gas as the fuel. The 'bulb' consists of a mantle and a reflector completes the lamp fitting which is attached to the fuel bottle by a flexible tube. These lamps produce a great deal of local heat and water vapour; the latter may have the effect of slowing down the drying out of the building. An alternative fuel to propane gas is butane gas but this fuel will not usually vaporise at temperatures below $-1\,^{\circ}C$.

Whichever method of illumination is used on a construction site it is always advisable to remember the axiom 'a workman can only be safe and work well when he can see where he is going and what he is doing'.

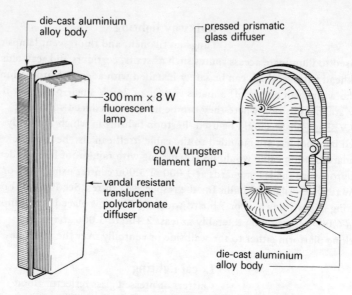

die-cast aluminium alloy body

pressed prismatic glass diffuser

300 mm × 8 W fluorescent lamp

60 W tungsten filament lamp

vandal resistant translucent polycarbonate diffuser

die-cast aluminium alloy body

Ceiling or wall mounted bulkhead lamp fittings

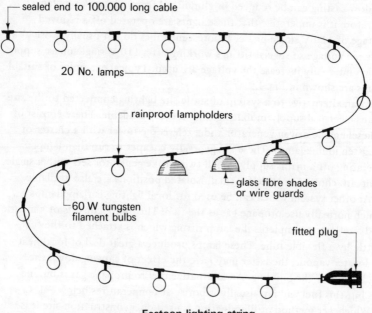

sealed end to 100.000 long cable

20 No. lamps

rainproof lampholders

glass fibre shades or wire guards

60 W tungsten filament bulbs

fitted plug

Festoon lighting string

Fig I.8 Typical walkway lighting fittings

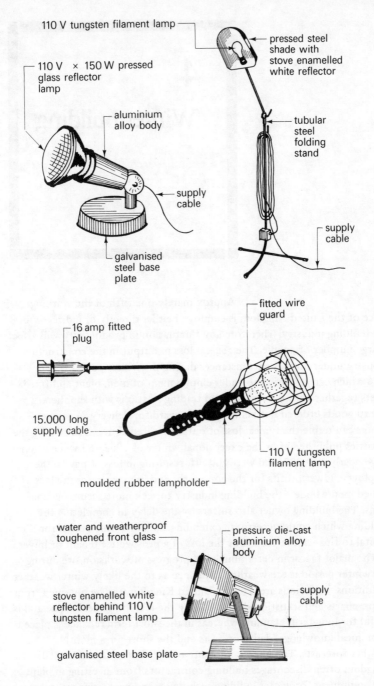

110 V tungsten filament lamp

pressed steel shade with stove enamelled white reflector

110 V × 150 W pressed glass reflector lamp

aluminium alloy body

tubular steel folding stand

supply cable

supply cable

galvanised steel base plate

fitted wire guard

16 amp fitted plug

15.000 long supply cable

110 V tungsten filament lamp

moulded rubber lampholder

water and weatherproof toughened front glass

pressure die-cast aluminium alloy body

supply cable

stove enamelled white reflector behind 110 V tungsten filament lamp

galvanised steel base plate

Fig I.9 Local lighting — suitable lamps and fittings

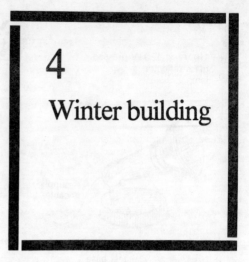

4
Winter building

Approximately one-fifth of the working force of the United Kingdom is employed either directly or indirectly by the building industry. Therefore any fluctuation in productivity will affect a large number of people. The general loss in output in the construction industry under normal circumstances during the winter period is about 10% which can result in the under-employment of men, plant and fixed assets together with the loss of good trading relations with suppliers due to goods ordered not being called forward for delivery. A severe winter can treble the typical loss of output quoted above as shown by the statistics published after the exceptional winter of 1962–63 when at one stage nearly 160 000 men were laid off, resulting in loss of pay to the employee, lower profits for the contractor and in many cases the loss of skilled men who left the building industry to seek more secure occupations. The building owner also suffers by the delay in completing the building which could necessitate extending the borrowing period for the capital to finance the project or the loss of a prospective tenant or buyer.

The major factor in determining the progress of works on site during the winter period is the weather. Guidance as to the likely winter weather conditions for various areas of the United Kingdom can be obtained from maps, charts and statistical data issued by the meteorological office and is useful for long-term planning whereas in the short term reliance is placed upon local knowledge, daily forecasts and the short-term monthly weather forecasts. The uncertain nature of the climate in the United Kingdom often discourages building contractors from investing in plant and equipment for winter building techniques and protective measures which may prove to be unnecessary. The contractor must therefore assess the total cost of possible delays against the capital outlay required for plant and equipment to enable him to maintain full or near full production during the winter period.

EFFECTS OF WEATHER

Weather conditions which can have a delaying effect on building activities are rain, high winds, low temperatures, snow and poor daylight levels; the worst effects obviously occur when more than one of the above conditions occur at the same time.

Rain: affects site access and movement which in turn increases site hazards, particularly those associated with excavations and earth moving works. It also causes discomfort to operatives thus reducing their productivity rate. Delays with most external operations such as bricklaying and concreting are usually experienced particularly during periods of heavy rainfall. Damage can be caused to unprotected materials stored on site and in many cases to newly fixed materials or finished surfaces. The higher moisture content of the atmosphere will also delay the drying out of buildings. If high winds and rain occur together, rain penetration and site hazards are considerably increased.

High winds: apart from the discomfort felt by operatives, high winds can also make activities such as frame erection and sheet cladding fixing very hazardous. They can also limit the operations that can be carried out by certain items of plant such as tower cranes and suspended cradles. Positive and negative wind pressures can also cause damage to partially fixed claddings, incomplete structures and materials stored on site.

Low temperatures: as the air temperature approaches freezing point many site activities are slowed down. These include excavating, bricklaying, concreting, plastering and painting, until they cease altogether at sub-zero temperatures; also mechanical plant can be difficult to start, while stockpiles of materials can become frozen and difficult to move. General movement and circulation around the site becomes hazardous, creating with the low temperatures general discomfort and danger for site personnel. When high winds are experienced with low temperatures they will aggravate the above-mentioned effects.

Snow: this is one of the most variable factors in British weather ranging from an average of five days a year on which snow falls on low ground in the extreme south-west to 35 days in the north-east of Scotland. Snow will impair the movement of labour, plant and materials as well as create uncomfortable working conditions. Externally stored materials will become covered with a layer of snow, making identification difficult in some cases. This blanket of snow will also add to the load to be carried by all horizontal surfaces. High winds encountered with falling snow can cause drifting which could increase the site hazards, personal discomfort and decrease general movement around the site.

It should be appreciated that the adverse conditions described above could have an adverse effect on site productivity even though they are not present on the actual site by delaying the movement of materials to the site from suppliers outside the immediate vicinity.

WINTER BUILDING TECHNIQUES
The major aim of any winter building method or technique is to maintain an acceptable rate of productivity. Inclement weather conditions can have a very quick reaction on the transportation aspect of site operations, movement of vehicles around the site and, indeed, off the site, which will be impaired or even brought to a complete standstill unless firm access roads or routes are provided, maintained and kept free of snow. These access roads should extend right up to the discharge points to avoid the need for unnecessary double handling of materials. If the access roads and hardstandings form part of the contract and are suitable, these could be constructed at an early stage in the contract before the winter period. If the permanent road system is not suitable in layout for contractural purposes, temporary roads of bulk timbers, timber or concrete sleepers, compacted hardcore or proprietary metal tracks could be laid.

Frozen ground can present problems with all excavating activities. Most excavating plant can operate in frozen ground up to a depth of 300 mm but at a reduced rate of output; this is particularly true when using machines having a small bucket capacity. Prevention is always better than cure. Therefore if frost is anticipated it is a wise precaution to protect the areas to be excavated by covering with straw mats enclosed in a polythene envelope, insulating quilts of mineral wool or glass fibre incorporating steam lines for severe conditions. Similar precautions can be taken in the case of newly excavated areas to prevent them freezing and giving rise to frost-heave conditions. If it is necessary to defrost ground to enable excavating works to be carried out, this can usually be achieved by using flame throwers, steam jet pipes or coils. Care must be taken to ensure that defrosting is complete and that precautions are taken to avoid subsequent re-freezing.

Water supplies should be laid below ground at such a depth so as to avoid the possibility of freezing, the actual depth will vary according to the locality of the site with a minimum depth of 750 mm for any area. If the water supply is temporary and above ground the pipes should be well lagged and laid to falls so that they can be drained at the end of the day through a drain cock incorporated into the service.

Electrical supplies can fail in adverse weather conditions due to the vulnerable parts such as contacts becoming affected by moisture, frost or ice. These components should be fully protected in a manner advised by an electrical contractor.

Items of plant which are normally kept uncovered on site — such as mixers, dumper trucks, bulldozers and generators — should be protected as recommended by the manufacturer to avoid morning starting problems. These precautions will include selecting and using the correct grades of oils, lubricants and antifreeze, also the covering of engines and electrical systems, draining radiators where necessary and parking wheeled or tracked vehicles on timber runners to prevent them freezing to the ground.

Men and materials will also need protection from adverse winter conditions if an acceptable level of production is to be maintained. Such protection can be of one or more of the following types:

1. Temporary shelters.
2. Framed enclosures.
3. Air-supported structures.
4. Protective clothing.

Temporary shelters: are the cheapest and simplest form of giving protection to the working areas consisting of a screen of reinforced or unreinforced polythene sheeting of suitable gauge fixed to the outside of the scaffold to form a windbreak. The sheeting must be attached firmly to the scaffold standards so that it does not flap or tear, a suitable method is shown in Fig. I.11. To gain the maximum amount of use and re-use out of the sheeting used to form the windbreaks the edges should be reinforced with a suitable adhesive tape incorporating metal eyelets at the tying positions. Eyelets can be made on site using a special kit or alternatively the sheet can be supplied with prepared edges.

Framed enclosures: consist of a purpose-made frame having a curved roof clad with a corrugated material and polythene sheeted sides or alternatively a frame enclosing the whole of the proposed structure can be constructed from standard tubular scaffolding components, see Fig. I.10. Framed enclosures should be clad from the windward end to avoid a build-up of pressure inside the enclosure. It is also advantageous to load the working platform before sheeting in the sides of the enclosure since loading at a later stage is more difficult. The frame must be rigid enough to take the extra loading of the coverings and any imposed loading such as

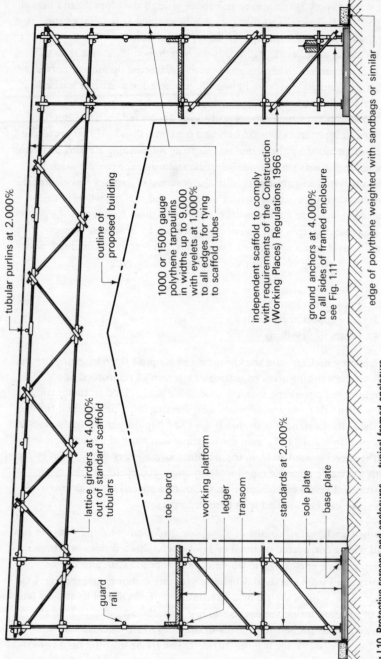

tubular purlins at 2.000%

lattice girders at 4.000% out of standard scaffold tubulars

outline of proposed building

1000 or 1500 gauge polythene tarpaulins in widths up to 9.000 with eyelets at 1.000% to all edges for tying to scaffold tubes

independent scaffold to comply with requirements of the Construction (Working Places) Regulations 1966

ground anchors at 4.000% to all sides of framed enclosure see Fig. 1.11

edge of polythene weighted with sandbags or similar

toe board

working platform

ledger

transom

standards at 2.000%

sole plate

base plate

guard rail

Fig 1.10 Protective screens and enclosures — typical framed enclosure

38

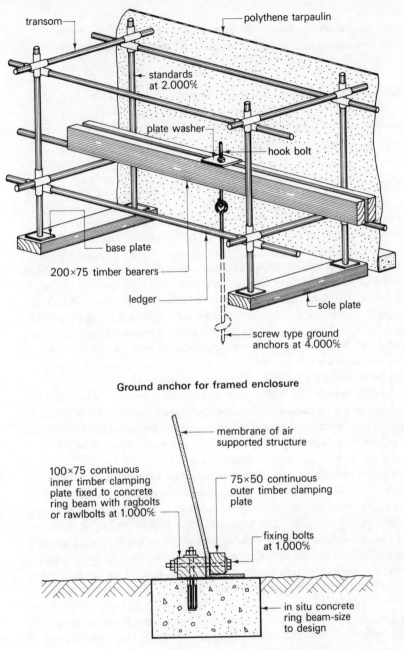

Ground anchor for framed enclosure

Ring beam anchorage for air supported structure

Fig I.11 Protective screens and enclosures — anchorages

wind loadings. Anchorage to the ground of the entire framing is also of great importance and this can be achieved by using a screw-type ground anchor as shown in Fig. I.11.

Air-supported structures: are sometimes called air domes are being increasingly used on building sites as a protective enclosure for works in progress and for covered material storage areas. Two forms are available:

1. Internally supported dome.
2. Air rib dome.

The internally supported dome is held up by internal air pressure acting against the covering membrane of some form of PVC coated nylon or rayon with access through an air lock or air curtain door, whereas in the air rib dome the membrane is supported by air-inflated ribs to which the covering membrane is attached. The usual shape for an air-supported structure is semi-cylindrical with rounded ends through which daylight can be introduced by having a translucent membrane over the crown of the structure. The advantages of air-supported structures are low cost, light in weight, re-usable, while only a small amount of labour is required to erect and dismantle them and with only a low internal pressure (approximately 150 N/m^2 above atmospheric pressure) workers inside are not affected. Disadvantages are the need to have at least one fan in continuous operation to maintain the internal air pressure, provision of an air lock or curtain entrance which will impede or restrict the general site circulation and the height limitation which is usually in the region of 45% of the overall span of the structure. The anchorage and sealing of the air-supported structure is also very important and this can be achieved by using a concrete ring beam as shown in Fig. I.11.

Protective clothing is another very important aspect of winter building techniques. Statistics show that the incidence of ailments usually associated with inclement weather such as rheumatism and bronchitis are some 20% higher in the building industry when compared with British industry as a whole. Labour costs on building sites are such that maximum utilisation of all labour resources in all weathers must be the ultimate aim and therefore capital expended in providing protective clothing can be a worthwhile investment. The ideal protective clothing consists of a suit in the form of jacket and trousers made from a lightweight polyurethane proofed nylon with a removable jacket liner for extreme conditions. A strong lightweight safety helmet and a strong pair of rubber boots with a good grip tread would complete the protective clothing outfit.

Heating equipment may be needed on a building site to offset the effects of cold weather on building operations, particularly where wet trades are involved when the materials used could become frozen at an early stage in the curing process producing an unacceptable component or member. The main types of heater in use are convectors, radiant heaters, salamanders and forced-air heaters using solid fuel, electricity, gas or oil as fuel. Solid fuel is generally considered to be impracticable because of its bulk and the problem of disposal of the ash residue. Electricity is usually too expensive in the context of building operations except for the operation of pumps and fans included in many of the heating appliances using in the main other fuels. Gas and oil are therefore the most used fuels and these must be considered in the context of capital and running costs.

Convector heaters using propane, paraffin or a light fuel with an output of 3 to 7 kW are only suitable for small volumes such as site huts and drying rooms where a suitable flue to conduct the fumes to the outside air can be installed.

Radiant heaters using propane are an alternative to the convector heater with the advantage of being mobile, making them suitable for site hut heating or for local heating where personnel are working.

Salamanders work on a similar principle to the convector heater using air currents but have a larger heating area giving outputs of 20 to 40 kW burning propane, paraffin or light fuel oil. They are suitable for heating enclosed areas where the heat output is contained and have the advantages of low capital costs, low running costs and that no flue is required.

Forced-air heaters have high outputs of 30 to 70 kW, are efficient, versatile and mobile. Two forms of heater are available: the direct forced-hot-air heater which discharges water vapour into the space being heated, and the vented forced-air heater where the combustion gases are discharged through a flue to the outside air. Both forms of heater require a fan operated by electricity, propane or a petrol engine. The usual fuels for the heater are propane, paraffin or light fuel oil which produces sufficient heat for general area heating of buildings under construction and for providing heat to several parts of a building using plastic ducting to distribute the hot air. When using a ducted distribution system it is essential that the air velocity at the nozzle is sufficient to distribute the hot air throughout the system.

The natural drying out of buildings after the wet trades is considerably slower in the winter months and therefore to meet production targets this drying out process needs the aid of suitable plant. The heaters described above are often used for this purpose but they are generally very inefficient. Effective drying out or the removal of excess moisture from a building requires the use of a de-humidifier and a vented space heater. Direct fired heaters are generally unsuitable since moisture is contained in the combustion gases discharged into the space being dried out. The de-humidifier is used in conjunction with a heater to eliminate the need of heating the cold air which would replace the warm moist air discharged to the outside if only a vented heater is employed. Two basic types of de-humidifier are available:

1. Refrigeration types where the air is drawn over refrigeration coils which lower the dewpoint causing the excess moisture in the air to condense and discharge into a container. According to the model used the extraction rate can be from 10 to 150 litres per day.
2. Chemical desiccant types where the moist air is drawn over trays or drums containing hygroscopic chemicals which attract the moisture from the air, the extracted moisture being transferred to a bucket or other suitable container. The advantages of these types are low capital and running costs and their overall efficiency with an extraction rate between 45 and 110 litres per day according to the model being used. The only real disadvantage is the need periodically to replenish the chemicals in the de-humidifier trays or drums.

The use of steam boilers and generators for defrosting plant and materials or heating water is also a possible winter building technique. The steam generator can be connected to steam coils which can be inserted into stockpiles of materials or alternatively the steam generator can be connected to a hand held lance.

The use of artificial lighting on sites to aid production and increase safety on building sites during the winter period when natural daylighting is often inadequate has been fully covered in the previous chapter.

WORKING WITH TRADITIONAL MATERIALS

Concrete: can be damaged by rain, sleet, snow, freezing temperatures and cooling winds before it has matured by slowing down the rate at which concrete hardens or by increasing the rate at which the water evaporates to an unacceptable level. The following precautions should ensure that no detrimental effects occur when mixing and placing in winter conditions:

1. Storage of cement under cover and in perfectly dry conditions to prevent air setting.
2. Defrosting of aggregates.
3. Minimum of delay between mixing and placing.
4. Minimum temperature of concrete when placed ideally should be 10°C.
5. Newly placed concrete to be kept at a temperature of more than 5°C for at least three days since the rate at which concrete sets

below this temperature is almost negligible. It may be necessary to employ the use of covers and/or heating to maintain this minimum temperature.

6. If special cements or additives such as calcium chloride in mass concrete are used the manufacturers' specifications should be strictly observed. It must be noted that calcium chloride should not be used in concrete mixes containing metal such as reinforcing bars.
7. Do not use antifreeze solutions.
8. Follow the recommendations of BS 8110 with regard to the stripping of formwork.

Brickwork: can be affected by the same climatic conditions as given above for concrete resulting in damage to the mortar joints and possibly spalling of the bricks. The following precautions should be taken when bricklaying under winter weather conditions:

1. Bricks to be kept reasonably dry.
2. Use a 1:1:6 mortar or a 1:5—6 cement:sand mortar with an air-entraining plasticiser to form microscopic air spaces which can act as expansion chambers for the minute ice particles which may form at low temperatures.
3. Cover up the brickwork upon completion with a protective insulating quilt or similar covering for at least three days.
4. During periods of heavy frost use a heated mortar by mixing with heated water (approximately $50°C$) to form a mortar with a temperature of between $15°$ and $25°C$.
5. Do not use additives such as calcium chloride since part of the solution could be absorbed by the absorbent bricks resulting in damage to the finished work.

Plastering: precautions which can be taken when plastering internally during cold periods are simple and consist of closing windows and doors or covering the openings with polythene or similar sheeting, maintaining an internal temperature above freezing and if necessary using heated (maximum $50°C$) mixing water. Lightweight plasters and aggregates are less susceptible to damage by frost than ordinary gypsum plasters and should be specified whenever possible. External plastering or rendering should only be carried out in dry weather when the air temperature is above freezing.

Timber: if the correct storage procedure of a covered but ventilated rack has been followed and the moisture content of the fixed joinery is maintained at the correct level few or no problems should be encountered when using this material during winter weather conditions.

If a contractor's output falls below his planned target figures then his fixed charges for overheads, site on-costs and plant costs are being wasted to a proportionate extent. His profit will also be less owing to the loss in turnover, therefore it is usually worth while expending an equivalent amount on winter building precautions and techniques to restore full production. In general terms a decrease of about 10% in productivity is experienced by builders during the winter period and by allocating 10% of his fixed charges and profit to winter building techniques a builder can bring his production back to normal. Cost analysis will show that the more exceptional the inclement weather becomes the greater is the financial reward for expended money on winter building techniques and precautions.

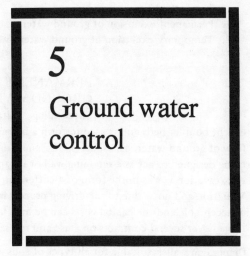

5
Ground water control

Ground water can be defined as water which is held temporarily in the soil above the level of the water table. Below the water table level is the subsoil water which is the result of the natural absorption by subsoils of the ground water. Both types of water can be effectively controlled by a variety of methods which have been designed either to exclude the water from a particular area or merely to lower the water table to give reasonably dry working conditions especially for excavation activities.

The extent to which water affects the stability and bearing capacity of a subsoil will depend upon the physical characteristics of the soil and in particular upon the particle size which ranges from the very fine particles of clay soils to the larger particles or boulders of some granular soils. The effect of the water on these particles is that of a lubricant enabling them to move when subjected to a force such as a foundation loading or simply causing them to flow by movement of the ground water. The number and disposition of the particles together with the amount of water present will determine the amount of movement which could take place. The finer particles will be displaced more easily than the larger particles which could create voids thus encouraging settlement of the larger particles.

The voids caused by excavation works encourage water to flow since the opposition to the ground water movement provided by the soil has been removed. In cases where the flow of water is likely an artificial opposition must be installed or the likelihood of water movement must be restricted by geotechnical processes. These processes can be broadly classified into one of two groups:

1. Permanent exclusion of ground water.
2. Temporary exclusion of ground water by lowering the water table.

PERMANENT EXCLUSION OF GROUND WATER

Sheet piling: suitable for all types of soils except boulder beds and is used to form a barrier or cut-off wall to the flow of ground water. The sheet piling can be of a permanent nature, being designed to act as a retaining wall or it can be a temporary enclosure to excavation works in the form of a cofferdam (for details see Chapter 5). Vibration and noise due to the driving process may render this method as unacceptable and the capital costs can be high unless they can be apportioned over several contracts on a use and re-use basis.

Diaphragm walls: suitable for all types of soils and are usually of *in situ* reinforced concrete installed using the bentonite slurry method (see Chapter 1, Volume 3 for details). This form of diaphragm wall has the advantages of low installation noise and vibration, can be used in restricted spaces and can be installed close to existing foundations. Generally, unless the diaphragm wall forms part of the permanent structure, this method is uneconomic.

Slurry trench cut-off: these are non-structural thin cast *in situ* unreinforced diaphragm walls suitable for subsoils of silts, sands and gravels. They can be used on sites where there is sufficient space to enclose the excavation area with a cut-off wall of this nature sited so that there is sufficient earth remaining between the wall and the excavation to give the screen or diaphragm wall support. Provided adequate support is given these walls are rapidly installed and are cheaper than the structural version.

Thin-grouted membrane: alternative method to the slurry trench cut-off wall when used in silt and sand subsoils, they are also suitable for installation in very permeable soils and made up ground where bentonite methods are unsuitable. Like the previous example ample earth support is required for this non-structural cut-off wall. The common method of formation is to drive into the ground a series of touching universal beam or column sections, sheet pile sections or alternatively small steel box sections to the required depth. A grout injection pipe is fixed to the web or face of the section and this is connected, by means of a flexible pipe, to a grout pump at ground level. As the sections are withdrawn the void created is filled with cement grout to form the thin membrane − see Fig. I.12.

Contiguous piling: an alternative method to the reinforced concrete

46

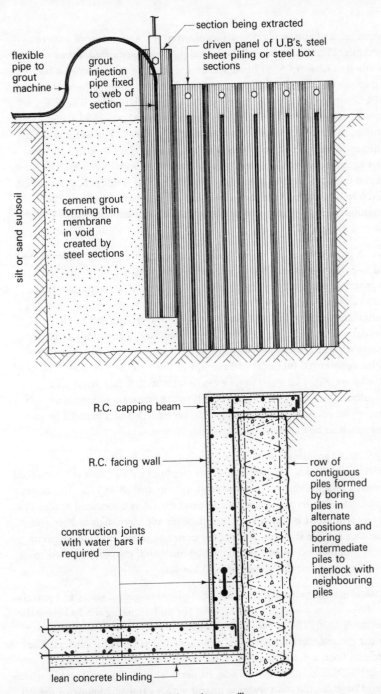

flexible pipe to grout machine

grout injection pipe fixed to web of section

section being extracted

driven panel of U.B's, steel sheet piling or steel box sections

cement grout forming thin membrane in void created by steel sections

silt or sand subsoil

R.C. capping beam

R.C. facing wall

construction joints with water bars if required

row of contiguous piles formed by boring piles in alternate positions and boring intermediate piles to interlock with neighbouring piles

lean concrete blinding

Fig I.12 Thin grouted membrane and contiguous piling

47

diaphragm wall consisting of a series of interlocking reinforced concrete-bored piles. The formation of the bored piles can be carried out as described in Chapter 5, Volume 3 ensuring that the piles interlock for their entire length. This will require special cutting tools to form the key in the alternate piles for the interlocking intermediate piles. The pile diameter selected will be determined by the strength required after completion of the excavations to one side of the wall. The usual range of diameters used is between 300 and 600 mm. Contiguous piling can be faced with a reinforced rendering or covered with a mesh reinforcement sprayed with concrete to give a smooth finish. This latter process is called Shotcrete or Gunite. An alternative method is to cast in front of the contiguous piling a reinforced wall terminating in a capping beam to the piles — see Fig. I.12.

Cement grouts: in common with all grouting methods cement grouts are used to form a 'curtain' in soils which have high permeability making temporary exclusion pumping methods uneconomic. Cement grouts are used in fissured and jointed rock stratas and are injected into the ground through a series of grouting holes bored into the ground in lines with secondary intermediate borehole lines if necessary. The grout can be a mixture of neat cement and water, cement and sand up to a ratio of 1 : 4 or PFA (pulverised fuel ash) and cement in the ratio of 1 : 1 with 2 parts of water by weight. The usual practice is to start with a thin grout and gradually reduce the water : cement ratio as the process proceeds to increase the viscosity of the mixture. To be effective this form of treatment needs to be extensive.

Clay/cement grouting: suitable for sands and gravels where the soil particles are too small for cement grout treatment. The grout is introduced by means of a sleeve grout pipe which limits its spread and like the cement grouting the equipment is simple and can be used in a confined space. The clay/cement grout is basically bentonite with additives such as Portland cement or soluble silicates to form the permanent barrier. One disadvantage of this method is that at least 4.000 of natural cover is required to provide support for the non-structural barrier.

Chemical grouting: suitable for use in medium-to-coarse sands and gravels to stabilise the soil and can also be used for underpinning works below the water table level. The chemicals are usually mixed prior to their injection into the ground through injection pipes inserted at 600 mm centres. The chemicals form a permanent gel or sol in the earth which increases the strength of the soil and also reduces its permeability. This method in which a liquid base diluted with water is mixed with a catalyst to control the gel setting time before being injected into the ground is called the one-shot method. An alternative two-shot method can be used and is carried out by injecting the first chemical (usually sodium silicate) into the ground and immediately afterwards injecting the second chemical (calcium chloride) to form a 'silica-gel'. The reaction of the two chemicals is immediate, whereas in the one-shot method the reaction of the chemicals can be delayed to allow for full penetration of the subsoil which will in turn allow a wider spacing of the boreholes. One main disadvantage of chemical grouting is the need for at least 2.000 of natural cover.

Resin grouts: suitable for silty fine sands or for use in conjunction with clay/cement grouts for treating fine strata but like the chemical grouts described above they can be costly unless used on large works. Resin grouts are similar in application to the chemical grouts but having a low viscosity this enables them to penetrate the fine sands unsuitable for chemical grouting applications.

Bituminous grouts: suitable for injection into fine sands to decrease the permeability of the soil but they will not increase the strength of the soil and are therefore unsuitable for underpinning work.

Grout injection: grouts of all kinds are usually injected into the subsoil by pumping in the mixture at high pressure through tubes placed at the appropriate centres according to the solution being used and/or the soil type. Soil investigation techniques will reveal the information required to enable the engineer to decide upon the pattern and spacing of the grout holes which can be drilled with pneumatic tools or tipped drills. The pressure needed to ensure a satisfactory penetration of the subsoil will depend upon the soil conditions and results required but is usually within the range of 1 N/mm^2 for fine soils to 7 N/mm^2 for cement grouting in fissured and jointed rock stratas.

Freezing: suitable method for all types of subsoils with a moisture content in excess of 8% of the voids. The basic principle is to insert freezing tubes into the ground and circulate a freezing solution around the tubes to form ice in the voids, thus creating a wall of ice to act as the impermeable barrier. This method will give the soil temporary extra mechanical strength but there is a slight risk of ground heave particularly when operating in clays and silts. The circulating solution can be a brine of magnesium chloride or calcium chloride at a temperature of between $-15°$ and $-25°$C which would take between 10 to 17 days to produce an ice wall 1.000 thick according to the type of subsoil. For works of short duration where quick freezing is required the more expensive liquid nitrogen can be used as the circulating medium. A typical freezing arrangement is shown in

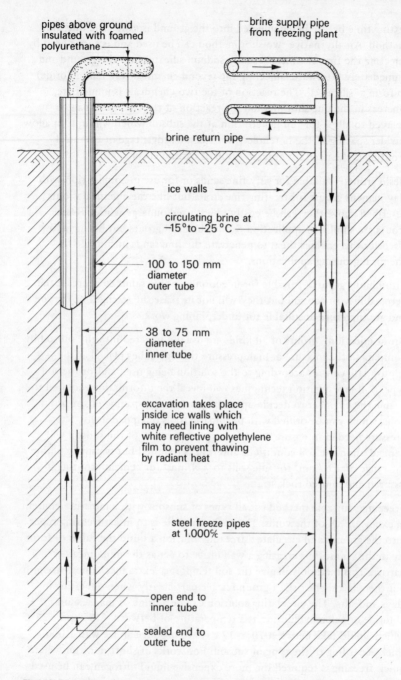

pipes above ground insulated with foamed polyurethane

brine supply pipe from freezing plant

brine return pipe

ice walls

circulating brine at –15° to –25 °C

100 to 150 mm diameter outer tube

38 to 75 mm diameter inner tube

excavation takes place inside ice walls which may need lining with white reflective polyethylene film to prevent thawing by radiant heat

steel freeze pipes at 1.000%

open end to inner tube

sealed end to outer tube

Fig I.13 Exclusion of ground water by freezing

Fig. I.13. Freezing methods of soil stabilisation are especially suitable for excavating deep shafts and driving tunnels.

TEMPORARY EXCLUSION OF GROUND WATER

Sump pumping: suitable for most subsoils and in particular gravels and coarse sands when working in open shallow excavations. The sump or water collection pit should be excavated below the formation level of the excavation and preferably sited in a corner position to reduce to a minimum the soil movement due to settlement which is a possibility with this method. Open sump pumping is usually limited to a maximum depth of 7.500 due to the limitations of suction lift of most pumps. An alternative method to the open sump pumping is the jetted sump which will achieve the same objective and will also prevent the soil movement. In this method a metal tube is jetted into the ground and the void created is filled with a sand media, a disposable hose and strainer as shown in Fig. I.14.

Wellpoint systems: popular method for water lowering in non-cohesive soils up to a depth of between 5.000 and 6.000. To dewater an area beyond this depth requires a multi-stage installation — see Fig. I.16. The basic principle is to water jet into the ground a number of small diameter wells which are connected to a header pipe which is attached to a vacuum pump — see Fig. I.15. Wellpoint systems can be installed with the header pipe acting as a ring main enclosing the area to be excavated. The header pipe should be connected to two pumps, the first for actual pumping operations and the second as a standby pump since it is essential to keep the system fully operational to avoid collapse of the excavation should a pump failure occur. The alternative system is the progressive line arrangement where the header pipe is placed alongside a trench or similar excavation to one side or both sides according to the width of the excavation. A pump is connected to a predetermined length of header pipe and further well points are jetted in ahead of the excavation works. As the work including backfilling is completed the redundant well points are removed and the header pipe is moved forwards.

Shallow-bored wells: suitable for sandy gravels and water-bearing rocks and is similar in principle to wellpoint pumping but is more appropriate than the latter for installations which have to be pumped for several months since running costs are generally less. This method is subject to the same lift restrictions as wellpoint systems and can be arranged as a multi-stage system if the depth of lowering exceeds 5.000.

51

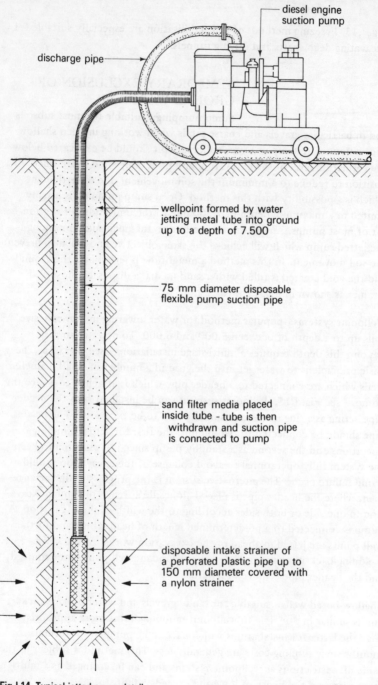

diesel engine
suction pump

discharge pipe

wellpoint formed by water
jetting metal tube into ground
up to a depth of 7.500

75 mm diameter disposable
flexible pump suction pipe

sand filter media placed
inside tube - tube is then
withdrawn and suction pipe
is connected to pump

disposable intake strainer of
a perforated plastic pipe up to
150 mm diameter covered with
a nylon strainer

Fig I.14 Typical jetted sump detail

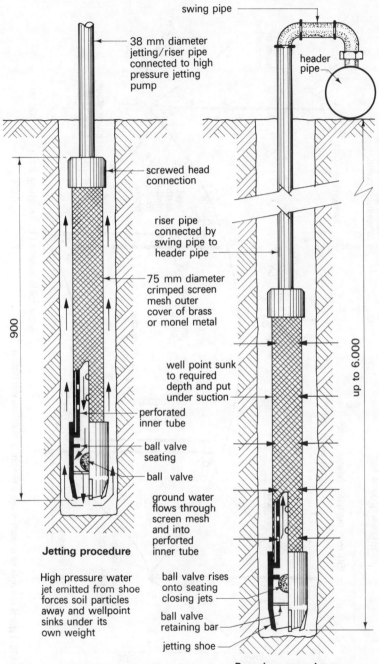

swing pipe

38 mm diameter jetting/riser pipe connected to high pressure jetting pump

header pipe

screwed head connection

riser pipe connected by swing pipe to header pipe

75 mm diameter crimped screen mesh outer cover of brass or monel metal

well point sunk to required depth and put under suction

perforated inner tube

ball valve seating

ball valve

ground water flows through screen mesh and into perforated inner tube

900

up to 6.000

Jetting procedure

High pressure water jet emitted from shoe forces soil particles away and wellpoint sinks under its own weight

ball valve rises onto seating closing jets

ball valve retaining bar

jetting shoe

Pumping procedure

Fig I.15 Typical wellpoint installation details

53

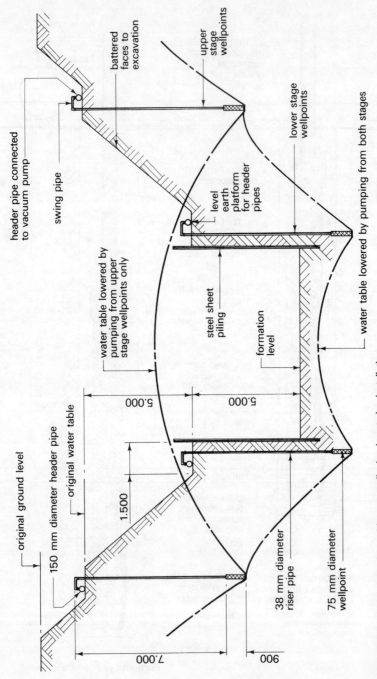

Fig I.16 Typical example of a multi-stage wellpoint dewatering installation

original ground level

header pipe connected to vacuum pump

swing pipe

battered faces to excavation

upper stage wellpoints

150 mm diameter header pipe

original water table

water table lowered by pumping from upper stage wellpoints only

lower stage wellpoints

level earth platform for header pipes

steel sheet piling

formation level

water table lowered by pumping from both stages

1.500

5.000

5.000

7.000

900

38 mm diameter riser pipe

75 mm diameter wellpoint

54

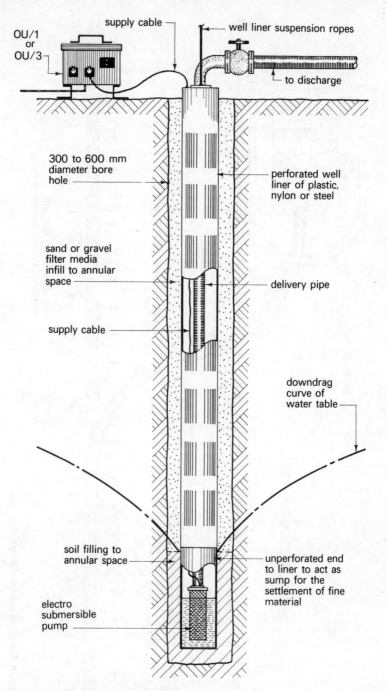

OU/1 or OU/3

supply cable

well liner suspension ropes

to discharge

300 to 600 mm diameter bore hole

perforated well liner of plastic, nylon or steel

sand or gravel filter media infill to annular space

delivery pipe

supply cable

downdrag curve of water table

soil filling to annular space

unperforated end to liner to act as sump for the settlement of fine material

electro submersible pump

Fig I.17 Typical deep-bored well details

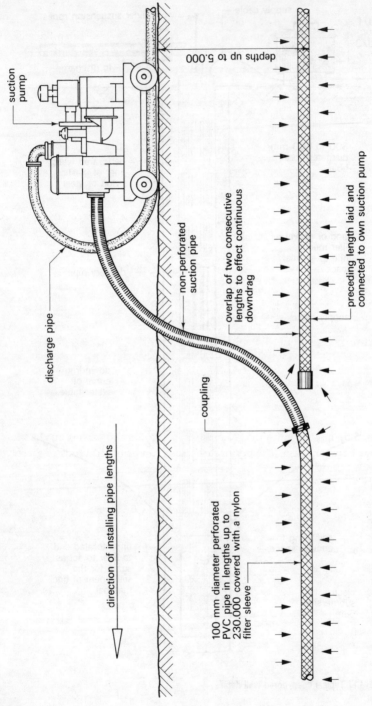

suction pump

discharge pipe

non-perforated suction pipe

coupling

direction of installing pipe lengths

depths up to 5.000

overlap of two consecutive lengths to effect continuous downdrag

preceding length laid and connected to own suction pump

100 mm diameter perforated PVC pipe in lengths up to 230.000 covered with a nylon filter sleeve

Fig I.18 Horizontal system of dewatering

Deep-bored wells: can be used as an alternative to a multi-stage wellpoint installation where the ground water needs to be lowered to a depth greater than 9.000. The wells are formed by sinking a 300 to 600 mm diameter steel lining tube into the ground to the required depth and at spacings to suit the subsoil being dewatered. This bore hole allows a perforated well liner to be installed with an electro submersible pump to extract the water. The annular space is filled with a suitable media such as sand and gravel as the outer steel lining tube is removed — see Fig. I.17.

Horizontal ground water control: the pumping methods described above all work on a completely vertical system. An alternative is the horizontal system of dewatering which consists of installing into the ground a 100 mm diameter PVC perforated suction pipe covered with a nylon filter sleeve to prevent the infiltration of fine particles. The pipe is installed using a special machine which excavates a narrow trench, lays the pipe and backfills the excavation in one operation at speeds up to 180 m per hour with a maximum depth of 5.000. Under average conditions a single pump can handle approximately 230 m of pipe run; for distances in excess of the pumping length an overlap of consecutive pipe lengths of up to 4.000 is required — see Fig. I.18.

Electro-osmosis: an uncommon and costly method which can be used for dewatering cohesive soils such as silts and clays where other pumping methods would not be adequate. This method works on the principle that soil particles carry a negative charge which attracts the positively charged ends of the water molecules creating a balanced state; if this balance is disturbed the water will flow. The disturbance of this natural balance is created by inserting into the ground two electrodes and passing an electric charge between them. The positive electrode can be of steel rods or sheet piling which will act as the anode and a wellpoint is installed to act as the cathode or negative electrode. When an electric current is passed between the anode and cathode it causes the positively charged water molecules to flow to the wellpoint (cathode) where it is collected and pumped away to a discharge point. The power consumption for this method can vary from 1 kW per m^3 for large excavations up to 12 kW per m^3 of soil dewatered for small excavations which will generally make this method uneconomic on running costs alone.

6
Cofferdams and caissons

A cofferdam may be defined as a temporary box structure constructed in earth or water to exclude soil and/or water from a construction area. It is usually formed to enable the formation of foundations to be carried out in safe working conditions. It is common practice to use interlocking steel trench sheeting or steel sheet piling to form the cofferdam but any material which will fulfil the same function can be used, including timber piles, precast concrete piles, earth-filled crib walls and banks of soil and rock. It must be clearly understood that to be economic and effective cofferdams must be structurally designed and such calculations are usually covered in the structural design syllabus of a typical course of study and have therefore been omitted from this text.

SHEET PILE COFFERDAMS

Cofferdams constructed from steel sheet piles or steel trench sheeting can be considered under two headings:

1. Single-skin cofferdams.
2. Double-skin cofferdams.

Single-skin cofferdams: these consist of a suitably supported single enclosing row of trench sheeting or sheet piles forming an almost completely watertight box. Trench sheeting could be considered for light loadings up to an excavation depth of 3.000 below the existing soil or water level whereas sheet piles are usually suitable for excavation depths of up to 15.000. The small amount of seepage which will occur through the interlocking joints must not be in excess of that which can be comfortably handled by a pump, or alternatively the joints can be sealed by caulking with asbestos rope, suitable mastics or a bitumastic compound.

Single-skin cofferdams constructed to act as cantilevers are possible in all soils but the maximum amount of excavation height will be low relative to the required penetration of the toe of the pile and this is particularly true in cohesive soils. Most cofferdams are therefore either braced and strutted or anchored using tie rods or ground anchors. Standard structural steel sections or structural timber members can be used to form the support system but generally timber is only economically suitable for low loadings. The total amount of timber required to brace a cofferdam adequately would be in the region of 0.25 to 0.3 m^3 per tonne of steel sheet piling used whereas the total weight of steel bracing would be in the region of 30 to 60% of the total weight of sheet piling used to form the cofferdam. Typical cofferdam support arrangements are shown in Figs. I.19 and 20. Single-skin cofferdams that are circular in plan can also be constructed using ring beams of concrete or steel to act as bracing without the need for strutting. Diameters up to 36.000 are economically possible using this method.

Cofferdams constructed in water, particularly those being erected in tidal waters, should be fitted with sluice gates to act as a precaution against unanticipated weaknesses in the arrangement and in the case of tidal waters to enable the water levels on both sides of the dam to be equalised during construction and before final closure. Special piles with an integral sluice gate forming a 200 mm wide x 400 mm deep opening are available. Alternatively a suitable gate can be formed by cutting a pair of piles and fitting them with a top-operated screw gear so that they can be raised to form an opening of any desired depth.

Double-skin cofferdams: these are self-supporting gravity structures constructed by using two parallel rows of piles with a filling material placed in the void created. Gravity-type cofferdams can also be formed by using straight-web sheet pile sections arranged as a cellular construction — see Figs. I.21 and 22.

The stability of these forms of cofferdam depends upon the design and arrangement of the sheet piling and upon the nature of the filling material. The inner wall of a double-skin cofferdam is designed as a retaining wall which is suitably driven into the sub-strata whereas the outer wall acts primarily as an anchor wall. The two parallel rows of piles are tied together with one or two rows of tie rods acting against external steel walings. Inner walls should have a series of low level weep holes to relieve the filling material of high water pressures and thus increase its shear resistance. For this reason the filling material selected should be capable of being well

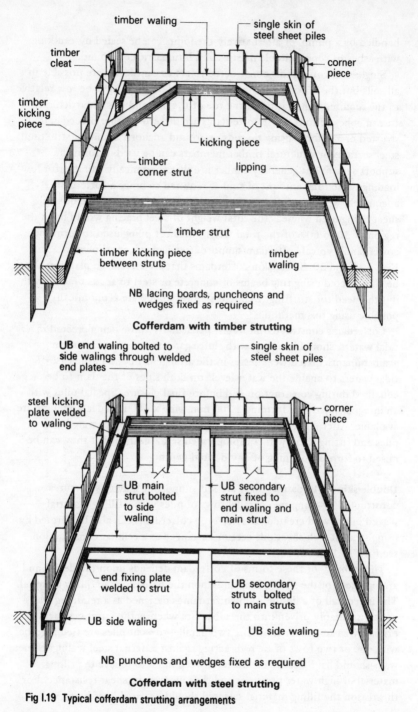

Cofferdam with timber strutting

NB lacing boards, puncheons and wedges fixed as required

Cofferdam with steel strutting

NB puncheons and wedges fixed as required

Fig I.19 Typical cofferdam strutting arrangements

60

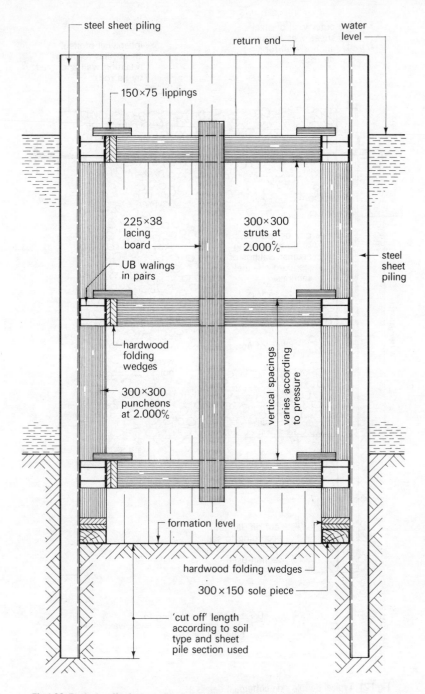

Fig I.20 Typical cofferdam section using timber and steel strutting

61

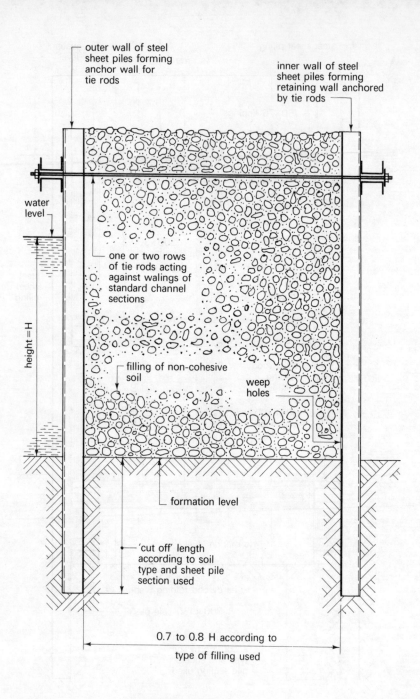

outer wall of steel
sheet piles forming
anchor wall for
tie rods

inner wall of steel
sheet piles forming
retaining wall anchored
by tie rods

water
level

one or two rows
of tie rods acting
against walings of
standard channel
sections

height = H

filling of non-cohesive
soil

weep
holes

formation level

'cut off' length
according to soil
type and sheet pile
section used

0.7 to 0.8 H according to

type of filling used

Fig I.21 Typical double skin cofferdam details

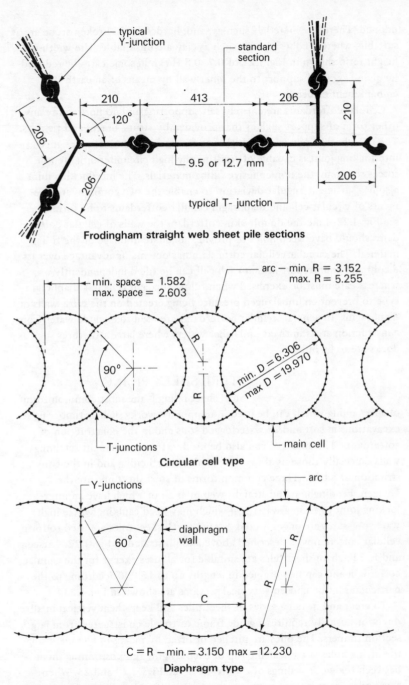

typical
Y-junction

standard
section

210

413

206

120°

206

206

206

206

210

9.5 or 12.7 mm

206

typical T- junction

Frodingham straight web sheet pile sections

min. space = 1.582
max. space = 2.603

arc — min. R = 3.152
max. R = 5.255

90°

R

R

min. D = 6.306
max D = 19.970

T-junctions

main cell

Circular cell type

Y-junctions

arc

60°

diaphragm
wall

R

R

R

C

C = R — min. = 3.150 max = 12.230
Diaphragm type

Fig I.22 Cellular cofferdam arrangements

drained. Therefore materials such as sand, hardcore and broken stone are suitable, whereas cohesive soils such as clay are unsuitable. The width-to-height ratio shown in Fig. I.21 of 0.7 : 0.8 H can in some cases be reduced by giving external support to the inner wall by means of an earth embankment or berm.

Cellular cofferdams are entirely self supporting and do not require any other form of support such as that provided by struts, braces and tie rods. The straight web pile with its high web strength and specially designed interlocking joint is capable of resisting the high circumferential tensile forces set up by the non-cohesive filling materials. The interlocking joint also has sufficient angular deviation to enable the two common arrangements of circular cell and diaphragm cellular cofferdams to be formed — see Fig. I.22. Like the double-skin cofferdam the walls of cellular cofferdams should have weep holes to provide adequate drainage of the filling material. The circular cellular cofferdam has one major advantage over its diaphragm counterpart in that each cell can be filled independently whereas care must be exercised when filling adjacent cells in a diaphragm type to prevent an unbalanced pressure being created on the cross-walls or diaphragms. In general, cellular cofferdams are used to exclude water from construction areas in rivers and other waters where large structures such as docks are to be built.

STEEL SHEET PILING

Steel sheet piling is the most common form of sheet piling which can be used in temporary works such as timbering to excavations in soft and/or waterlogged soils and in the construction of cofferdams. This material can also be used to form permanent retaining walls especially those used for river bank strengthening and in the construction of jetties. Three common forms of steel sheet pile are the Larseen, Frodingham and straight-web piles all of which have an interlocking joint to form a water seal which may need caulking where high water pressures are encountered. Straight-web sheet piles are used to form cellular cofferdams as described above and illustrated in Fig. I.22. Larseen and Frodingham sheet piles are suitable for all uses except for the cellular cofferdam and can be obtained in lengths up to 18.000 according to the particular section chosen — typical sections are shown in Fig. I.23.

To erect and install a series of sheet piles and keep them vertical in all directions usually requires a guide frame or trestle constructed from large section timbers. The piles are pitched or lifted by means of a crane, using the lifting holes sited near the top of each length, and positioning them between the guide walings of the trestle — see Figs. I.24 and 25. When sheet piles are being driven there is a tendency for them to creep or lean in

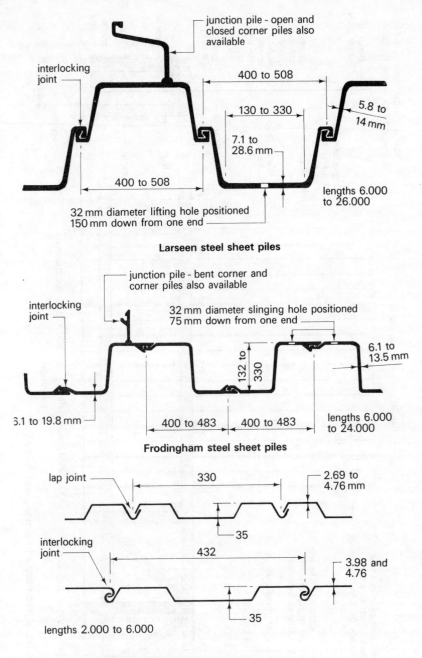

junction pile - open and closed corner piles also available

interlocking joint

400 to 508

130 to 330

7.1 to 28.6 mm

5.8 to 14 mm

400 to 508

32 mm diameter lifting hole positioned 150 mm down from one end

lengths 6.000 to 26.000

Larseen steel sheet piles

junction pile - bent corner and corner piles also available

interlocking joint

32 mm diameter slinging hole positioned 75 mm down from one end

132 to 330

6.1 to 13.5 mm

6.1 to 19.8 mm

400 to 483

400 to 483

lengths 6.000 to 24.000

Frodingham steel sheet piles

lap joint

330

2.69 to 4.76 mm

35

interlocking joint

432

3.98 and 4.76

35

lengths 2.000 to 6.000

Steel trench sheeting

Fig I.23 Typical steel sheet pile and trench sheeting sections

65

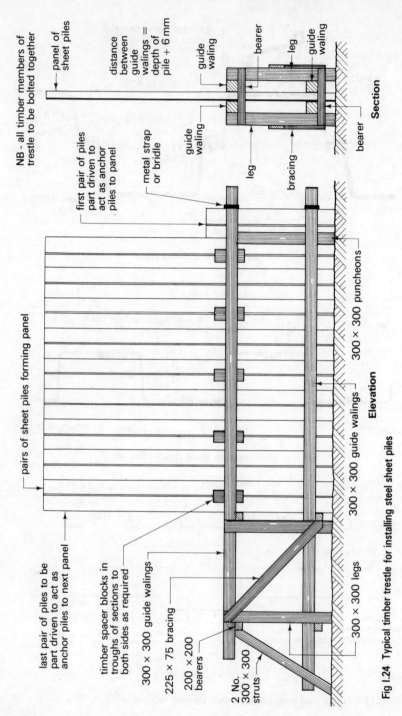

NB - all timber members of trestle to be bolted together

panel of sheet piles

distance between guide walings = depth of pile + 6 mm

guide waling

bearer

leg

guide waling

Section

guide waling

leg

bracing

bearer

first pair of piles part driven to act as anchor piles to panel

metal strap or bridle

pairs of sheet piles forming panel

300 × 300 puncheons

300 × 300 guide walings

Elevation

300 × 300 legs

last pair of piles to be part driven to act as anchor piles to next panel

timber spacer blocks in troughs of sections to both sides as required

300 × 300 guide walings

225 × 75 bracing

200 × 200 bearers

2 No. 300 × 300 struts

Fig I.24 Typical timber trestle for installing steel sheet piles

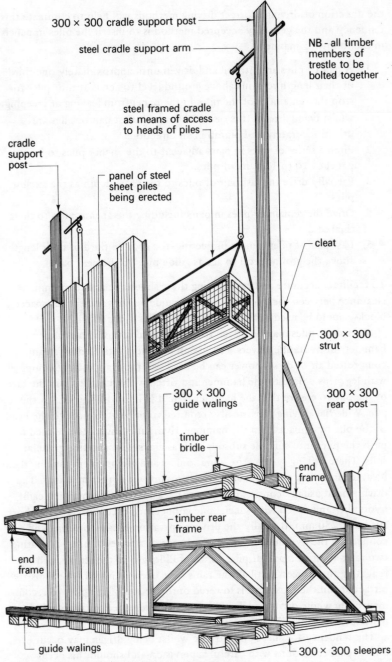

300 × 300 cradle support post

steel cradle support arm

NB - all timber members of trestle to be bolted together

steel framed cradle as means of access to heads of piles

cradle support post

panel of steel sheet piles being erected

cleat

300 × 300 strut

300 × 300 guide walings

300 × 300 rear post

timber bridle

end frame

timber rear frame

end frame

end frame

guide walings

300 × 300 sleepers

Fig I.25 Typical timber trestle for large steel sheet piles

67

the direction of driving. Correct driving methods will help to eliminate this tendency and the generally accepted method is to install the piles in panels in the following manner:

1. A pair of piles are pitched and driven until approximately one-third of their length remains above ground level to act as anchor piles to stop the remainder of the piles in the panel from leaning or creeping whilst being driven. It is essential that this first pair of piles are driven accurately and plumb in all directions.
2. Pitch a series of piles in pairs adjacent to the anchor piles to form a panel of 10 to 12 pairs of piles.
3. Partially drive the last pair of piles to the same depth as the anchor piles.
4. Drive the remaining piles in pairs including the anchor piles to their final set.
5. Last pair of piles remain projecting, for about a third of their length, above the ground level to act as guide piles to the next panel.

To facilitate accurate and easy driving there should be about a 6 mm clearance between the pile faces and the guide walings and timber spacer blocks should be used in the troughs of the piles — see Fig. I.24.

Steel sheet piles may be driven to the required set using percussion hammers or hydraulic drivers. Percussion hammers activated by steam, compressed air or diesel power can be used and these are usually equipped with leg grips bolted to the hammer and fitted with inserts to grip the face of the pile to ensure that the hammer is held in line with the axis of the pile. Wide flat driving caps are also required to prevent damage to the head of the pile by impact from the hammer. Hydraulic drivers can be used to push the piles into suitable subsoils such as clays, silts and fine granular soils. These driving systems are vibrationless and almost silent making them ideal for installing sheet piling in close proximity to other buildings. The driving head usually consists of a power pack crosshead containing eight hydraulic rams fitted with special pile connectors, each ram having a short stroke of 750 mm. Basically the driving operation entails lowering and connecting the hydraulic driver to the heads of a panel of piles, activating two rams to drive a pair of piles for the full length of the stroke, and repeating this process until all the rams have been activated. The rams are retracted and the crosshead is lowered onto the top of the piles to recommence the whole driving cycle. This process is repeated until the required penetration of the piles has been reached.

The tendency for sheet piles to lean whilst being driven may occur even under ideal conditions with careful supervision and should this occur immediate steps must be taken to correct the fault or it may get out of control. The following correction methods can be considered:

1. Attach a winch rope to the pair of piles being driven and exert a corrective pulling force as driving continues.
2. Attach a winch rope to the previously driven adjacent pair of piles and exert a corrective pulling force as driving continues.
3. Reposition the hammer towards the previously driven pair of piles to give an eccentric blow.
4. Combinations of any of the above methods.

If the above correction techniques are not suitable or effective it is possible to form tapered piles by making use of the tolerances within the interlocking joint by welding steel straps across the face of the pair of piles to hold them in tapered form. This technique is only suitable if the total amount of taper required does not exceed 50 mm in the length of the pile. To insert a special tapered pile within a panel it will of course be necessary to extract a pair of piles from the panel. Typical examples of correction techniques are shown in Fig. I.26.

Water jetting

In soft and silty subsoils the installation of steel sheet piles can be assisted by using high-pressure water jetting to the sides and toes of the piles. The jets should be positioned in the troughs of the piles and to both sides of the section. In very soft subsoils this method can sometimes be so effective that the assistance of a driving hammer is not required. To ensure that the pile finally penetrates into an undisturbed layer of subsoil the last few metres or so of driving should be carried out without jetting.

Access

Access in the form of a working platform to the heads of sheet piles is often necessary during installation to locate adjacent piles, position and attach the driving hammer and to carry out inspections of the work in progress. Suitable means of access are:

1. Independent scaffolding
2. Suspended cradle — see Fig. I.25
3. Mobile platform mounted on a hydraulic arm
4. Working platform in the form of a cage suspended from a mobile crane and hooked onto the pile heads — see Fig. I.26.

Extraction of piles

Piles which are used in temporary works should have well-greased interlocking joints to enable them to be extracted with ease. Inverted double-acting hammers can be used to extract sheet piles but these have been

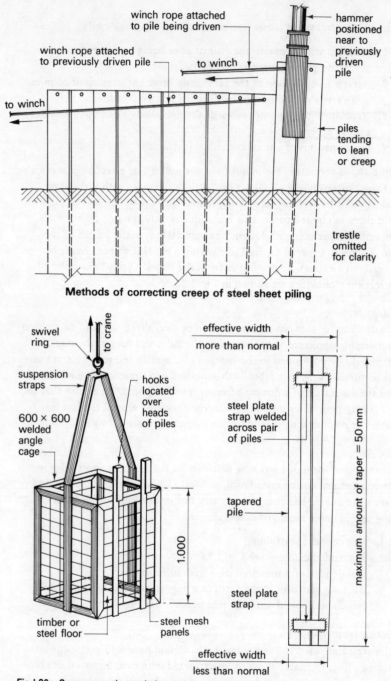

Methods of correcting creep of steel sheet piling

winch rope attached to pile being driven

winch rope attached to previously driven pile

to winch

hammer positioned near to previously driven pile

to winch

piles tending to lean or creep

trestle omitted for clarity

swivel ring

to crane

suspension straps

hooks located over heads of piles

600 × 600 welded angle cage

steel plate strap welded across pair of piles

1.000

tapered pile

timber or steel floor

steel mesh panels

steel plate strap

effective width more than normal

maximum amount of taper = 50 mm

effective width less than normal

Fig I.26 Creep correction techniques and access cage

generally superseded by specially designed sheet pile extractors. These usually consist of a compressed air- or steam-activated ram giving between 120 and 200 upward blows per minute which causes the jaws at the lower end to grip the pile and force it out of the ground.

CAISSONS

These are box-like structures which can be sunk through ground or water to install foundations or similar structures below the water line or table. They differ from cofferdams in that they usually become part of the finished foundation or structure and should be considered as an alternative to the temporary cofferdam if the working depth below the water level exceeds 18.000. The design and installation of the various types of caissons is usually the task of a specialist organisation but building contractors should have a fundamental knowledge of the different types and their uses.

There are four basic types of caisson in general use, namely:

1. Box caissons.
2. Open caissons.
3. Monolithic caissons.
4. Pneumatic or compressed air caissons.

Box caissons: these are prefabricated precast concrete boxes which are open at the top and closed at the bottom. They are usually constructed on land and designed to be launched and floated to the desired position where they are sunk onto a previously prepared dredged or rock blanket foundation — see Fig. I.27. If the bed strata is unsuitable for the above preparations it may be necessary to lay a concrete raft by using traditional cofferdam techniques onto which the caisson can be sunk. During installation it is imperative that precautions are taken to overcome the problems of floatation by flooding the void with water or adding kentledge to the caisson walls. The sides of the caisson will extend above the water line after it has been finally positioned providing a suitable shell for such structures as bridge piers, breakwaters and jetties. The void is filled with *in situ* concrete placed by pump, tremie pipe or crane and skip. Box caissons are suitable for situations where the bed conditions are such that it is not necessary to sink the caisson below the prepared bed level.

Open caissons: sometimes referred to as cylinder caissons because of their usual plan shape. They are of precast concrete and open at both top and bottom ends with a cutting edge to the bottom rim. They are suitable for installation in soft subsoils where the excavation can be carried out by conventional grabs enabling the caisson to sink under its own weight as the

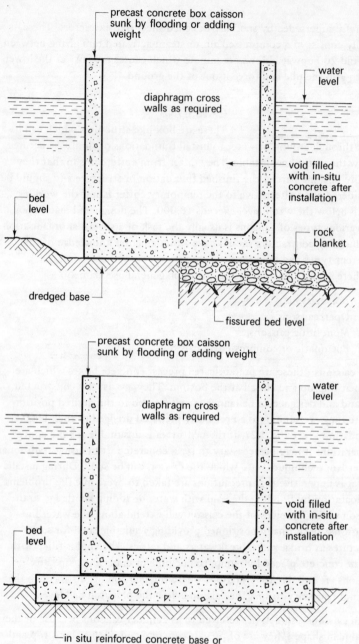

precast concrete box caisson
sunk by flooding or adding
weight

diaphragm cross
walls as required

water
level

void filled
with in-situ
concrete after
installation

rock
blanket

bed
level

dredged base

fissured bed level

precast concrete box caisson
sunk by flooding or adding weight

diaphragm cross
walls as required

water
level

void filled
with in-situ
concrete after
installation

bed
level

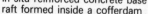

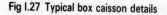

in situ reinforced concrete base or
raft formed inside a cofferdam

Fig I.27 Typical box caisson details

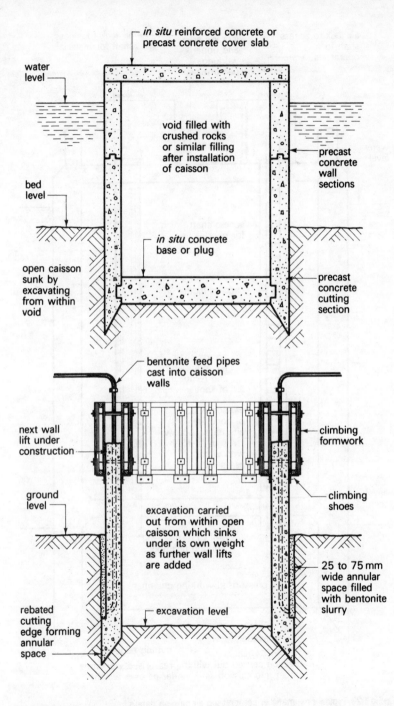

Fig 1.28 Typical open caisson details

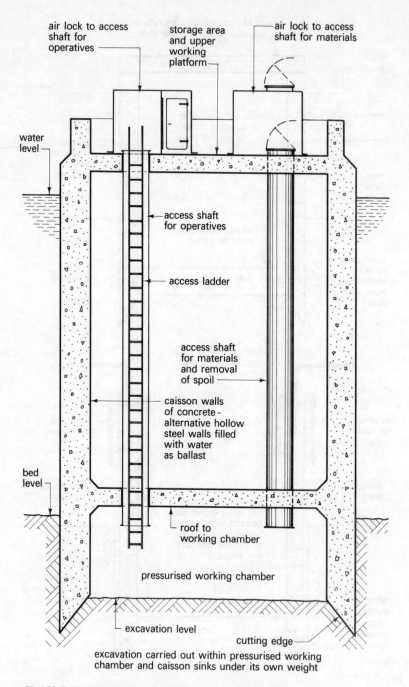

air lock to access shaft for operatives

storage area and upper working platform

air lock to access shaft for materials

water level

access shaft for operatives

access ladder

access shaft for materials and removal of spoil

caisson walls of concrete - alternative hollow steel walls filled with water as ballast

bed level

roof to working chamber

pressurised working chamber

excavation level

cutting edge

excavation carried out within pressurised working chamber and caisson sinks under its own weight

Fig I.29 Typical pneumatic or compressed air caisson details

excavation proceeds. These caissons can be completely or partially pre-formed whereas in the case of the latter further sections can be added or cast on as the structure sinks to the required depth. When the desired depth has been reached a concrete plug in the form of a slab is placed in the bottom by tremie pipe to prevent further ingress of water. The cell void can now be pumped dry and filled with crushed rocks or similar material if necessary to overcome floatation during further construction works.

Open caissons can also be installed in land if the subsoil conditions are suitable. The shoe or cutting edge is formed so that it is wider than the wall above to create an annular space some 75 to 100 mm wide into which a bentonite slurry can be pumped to act as a lubricant and thus reduce the skin friction to a minimum. Excavation is carried out by traditional means within the caisson void, the caisson sinking under its own weight. The excavation operation is usually carried out simultaneously with the construction of the caisson walls above ground level — see Fig. I.28.

Monolithic caissons: these are usually rectangular in plan and are divided into a number of voids or wells through which the excavation is carried out. They are similar to open caissons but have greater self weight and wall thickness, making them suitable for structures such as quays which may have to resist considerable impact forces in their final condition.

Pneumatic or compressed-air caissons: these are similar to open caissons except that there is an air-tight working chamber some 3.000 high at the cutting edge. They are used where difficult subsoils exist and where hand excavation in dry working conditions is necessary. The working chamber must be pressurised sufficiently to control the inflow of water and/or soil and at the same time provide safe working conditions for the operatives. The maximum safe working pressure is usually specified as 310 kN/m^2 which will limit the working depth of this type of caisson to about 28.000. When the required depth has been reached the floor of the working chamber can be sealed over with a 600 mm thick layer of well vibrated concrete. This is followed by further well-vibrated layers of concrete until only a small space remains which is pressure grouted to finally seal the working chamber. The access shafts are finally sealed with concrete some three to four days after sealing off the working chamber — for typical details see Fig. I.29.

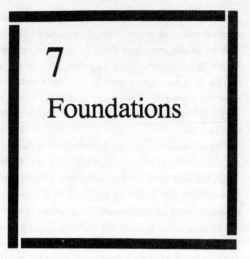

7
Foundations

Part II
Substructure

Loadings in buildings consist of the combined dead and imposed loads which exert a downward pressure upon the soil on which the structure is founded and this in turn promotes a reactive force in the form of an upward pressure from the soil. The structure is in effect sandwiched between these opposite pressures and the design of the building must be able to resist the resultant stresses set up within the structural members and the general building fabric. The supporting subsoil must be able to develop sufficient reactive force to give stability to the structure to prevent failure due to unequal settlement and to prevent failure of the subsoil due to shear. To enable a designer to select, design and detail a suitable foundation he must have adequate data regarding the nature of the soil on which the structure will be founded and this is normally obtained from a planned soil investigation programme.

SOIL INVESTIGATION

Soil investigation is specific in its requirements whereas site investigation is all embracing, taking into account such factors as topography, location of existing services, means of access and any local restrictions. Soil investigation is a means of obtaining data regarding the properties and characteristics of subsoils by providing samples for testing or providing a means of access for visual inspection. The actual data required and the amount of capital which can be reasonably expended on any soil investigation programme will depend upon the type of structure proposed and how much previous knowledge the designer has of a particular region or site.

The main methods of soil investigation can be enumerated as follows:

1. Trial pits — small contracts where foundation depths are not likely to exceed 3.000.
2. Boreholes — medium to large contracts with foundations up to 30.000 deep.

Trial pits: relatively cheap and easy method of obtaining soil data by enabling easy visual inspection of the soil strata in its natural condition. The pits can be hand or machine excavated to a plan size of 1.200 x 1.200 and spaced at centres to suit the scope of the investigation. A series of pits set out on a 20.000 grid would give a reasonable coverage of most sites. The pits need to be positioned so that the data obtained is truly representative of the actual conditions but not in such a position where their presence could have a detrimental effect on the proposed foundations. In very loose soils or soils having a high water table trial pits can prove to be uneconomical due to the need for pumps and/or timbering to keep the pits dry and accessible. The spoil removed will provide disturbed samples for testing purposes whereas undisturbed samples can be cut and extracted from the walls of the pit.

Boreholes: these enable disturbed or undisturbed samples to be removed for analysis and testing but undisturbed samples are sometimes difficult to obtain from soils other than rock or cohesive soils. The core diameter of the samples obtained vary from 100 to 200 mm according to the method employed in extracting the sample. Disturbed samples can be obtained by using a rotary flight auger or by percussion boring in a similar manner to the formation of small diameter bored piles using a tripod or shear leg rig. Undisturbed samples can be obtained from cohesive soils using 450 mm long x 100 mm diameter sampling tubes which are driven into the soil to collect the sample within itself; upon removal the tube is capped, labelled and sent off to a laboratory for testing. Undisturbed rock samples can be obtained by core drilling with diamond tipped drills where necessary.

CLASSIFICATION OF SOILS

Soils may be classified by any of the following methods:

1. Physical properties.
2. Geological origin.
3. Chemical composition.
4. Particle size.

It has been established that the physical properties of soils can be closely associated with their particle size both of which are of importance to the foundation engineer, architect or designer. All soils can be defined as being coarse-grained or fine-grained each resulting in different properties.

Coarse-grained soils: these would include sands and gravels having a low proportion of voids, negligible cohesion when dry, high permeability and slight compressibility, which takes place almost immediately upon the application of load.

Fine-grained soils: these include the cohesive silts and clays having a high proportion of voids, high cohesion, very low permeability and high compressibility which takes place slowly over a long period of time.

There are of course soils which can be classified in between the two extremes described above. BS 1377 deals with the methods of testing soils and divides particle sizes as follows:

Clay particles	less than 0.002 mm
Silt particles	between 0.002 and 0.06 mm
Sand particles	between 0.06 and 2 mm
Gravel particles	between 2 and 60 mm
Cobbles	between 60 and 200 mm

The silt, sand and gravel particles are also further subdivided into fine, medium and coarse with particle sizes lying between the extremes quoted above.

Fine-grained soils such as clays are difficult to classify positively by their particle size distribution alone and therefore use is made of the reaction of these soils to a change in their moisture content. If the moisture content is high the volume is consequently large and the soil is basically a suspension of clay particles in water. As the moisture content decreases the soil passes the liquid limit and becomes plastic. The liquid limit of a soil is defined as 'the moisture content at which a soil passes from the plastic state to the liquid state' and this limit can be determined by the test set out in BS 1377.

Further lowering of the moisture content will enable the soil to pass the plastic limit and begin to become a solid. The plastic limit of a soil is reached when a 20 g sample just fails to roll into a 3 mm diameter thread when rolled between the palm of the hand and a glass plate. When soils of this nature reach the solid state the volume tends to remain constant and any further decrease in moisture content will only alter the appearance and colour of the sample. It must be clearly understood that the change from one definable state to another is a gradual process and not a sudden change.

78

Shear strength of soils

The resistance which can be offered by a soil to the sliding of one portion over another or its shear strength is of importance to the designer since it can be used to calculate the bearing capacity of a soil and the pressure it can exert on such members as timbering in excavations. Resistance to shear in a soil under load depends mainly upon its particle composition. If a soil is granular in form the frictional resistance between the particles increases with the load applied and consequently its shear strength also increases with the magnitude of the applied load. Conversely clay particles being small develop no frictional resistance and therefore its shear strength will remain constant whatever the magnitude of the applied load. Intermediate soils such as sandy clays normally give only a slight increase in shear strength as the load is applied.

To ascertain the shear strength of a particular soil sample the triaxial compression test as described in BS 1377 is usually employed for cohesive soils. Non-cohesive soils can be tested in a shear box which consists of a split box into which the sample is placed and subjected to a standard vertical load whilst a horizontal load is applied to the lower half of the box until the sample shears.

Compressibility

Another important property of soils which must be ascertained before a final choice of foundation type and design can be made is compressibility and two factors must be taken into account:

1. Rate at which compression takes place.
2. Total amount of compression when full load is applied.

When dealing with non-cohesive soils such as sands and gravels the rate of compression will keep pace with the construction of the building and therefore when the structure is complete there should be no further settlement if the soil remains in the same state. A soil is compressed when loaded by the expulsion of air and/or water from the voids and by the natural rearrangement of the particles. In cohesive soils the voids are very often completely saturated with water which in itself is nearly incompressible and therefore compression of the soil can only take place by the water moving out of the voids thus allowing settlement of the particles. Expulsion of water from the voids within cohesive soils can occur but only at a very slow rate due mainly to the resistance offered by the plate-like particles of the soil through which it must flow. This gradual compressive movement of a soil is called consolidation. Uniform settlement will not normally cause undue damage to a structure but uneven settlement can cause progressive structural damage.

Stresses and pressures

The above comments on shear strength and compressibility clearly indicate that cohesive soils present the most serious problems when giving consideration to foundation choice and design. The two major conditions to be considered are:

1. Shearing stresses.
2. Vertical pressures.

Shearing stress: the maximum stress under a typical foundation carrying a uniformly distributed load will occur on a semi-circle whose radius is equal to half the width of the foundation and the isoshear line value will be equal to about one-third the applied pressure — see Fig. II.1. The magnitude of this maximum pressure should not exceed the shearing resistance value of the soil.

Vertical pressure: this acts within the mass of the soil upon which the structure is founded and should not be of such a magnitude as to cause unacceptable settlement of the structure. Vertical pressures can be represented on a drawing by connecting together points which have the same value forming what are termed pressure bulbs. Most pressure bulbs are plotted up to a value of 0.2 of the pressure per unit area which is considered to be the limit of pressure which could influence settlement of the structure. Typical pressure bulbs are shown in Figs. II.1 and 2. A comparison of these typical pressure bulbs will show that generally vertical pressure decreases with depth, the 0.2 value will occur at a lower level under strip foundations than under rafts, isolated bases and bases in close proximity to one another which form combined pressure bulbs. The pressure bulbs illustrated in Figs. II.1 and 2 are based on the soil being homogeneous throughout the depth under consideration. As in reality this is not always the case it is important that soil investigation is carried out at least to the depth of the theoretical pressure bulb. Great care must be taken where an underlying strata of highly compressible soil is encountered to ensure that these are not over stressed if cut by the anticipated pressure bulb.

Contact pressure

It is very often incorrectly assumed that a foundation which is uniformly loaded will result in a uniform contact pressure under the foundation. This would only be true if the foundation was completely flexible such as the bases to a pin jointed frame. The actual contact pressure under a foundation will be governed by the nature of the soil and the rigidity of the foundation, and since in practice most large structures have a rigid foundation the contact pressure distribution is not uniform. In cohesive soils there is a tendency for high stresses to occur at the edges which is usually reduced slightly by the yielding of the clay soil. Non-cohesive soils give rise to a parabolic contact pressure distribution with increasing edge pressures as the depth below ground level of the foundation increases. When selecting the basic foundation format consideration must be given to the concentration of the major loads over the position where the theoretical contact pressures are at a minimum to obtain a balanced distribution of contact pressure — see Fig. II.3.

Plastic failure

This is a form of failure which can occur in cohesive soils if the ultimate bearing capacity of the soil is reached or exceeded. As the load on a foundation is increased the stresses within the soil also increases until all resistance to settlement has been overcome. Plastic failure, which can be related to the shear strength of the soil, occurs when the lateral pressure being exerted by the wedge of relatively undisturbed soil immediately below the foundation causes a plastic shear failure to develop resulting in a heaving of the soil at the sides of the foundation moving along a slip circle or plane. In practice this movement tends to occur on one side of the building, causing it to tilt and settle — see Fig. II.4. Plastic failure is likely to happen when the pressure applied by the foundation is approximately six times the shear strength of the soil.

FOUNDATION TYPES

There are many ways in which foundations can be classified but one of the most common methods is by form resulting in five basic types thus:

1. *Strip foundations* — light loadings particularly in domestic buildings — see Chapter 3, Volume 1. Heavier loadings can sometimes be founded on a reinforced concrete strip foundation as described and illustrated in Chapter 7, Volume 2.
2. *Raft foundations* — light loadings, average loadings on soils with low bearing capacities and structures having a basement storey — see Chapter 7, Volume 2.
3. *Pad or isolated foundations* — common method of providing the foundation for columns of framed structures and for the supporting members of portal frames — see Chapter 7, Volume 2 and Chapter 7, Volume 3.

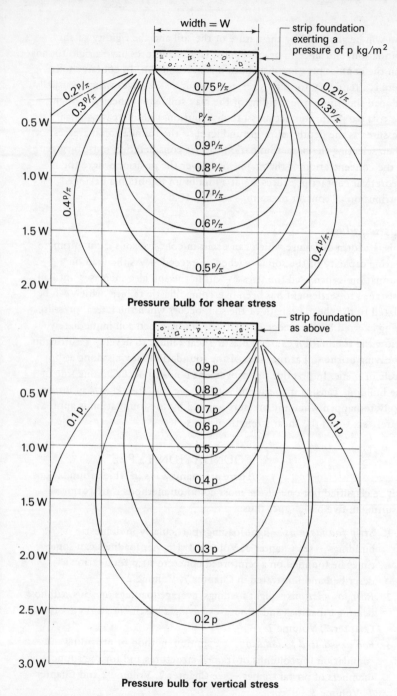

Fig II.1 Strip foundations — typical pressure bulbs

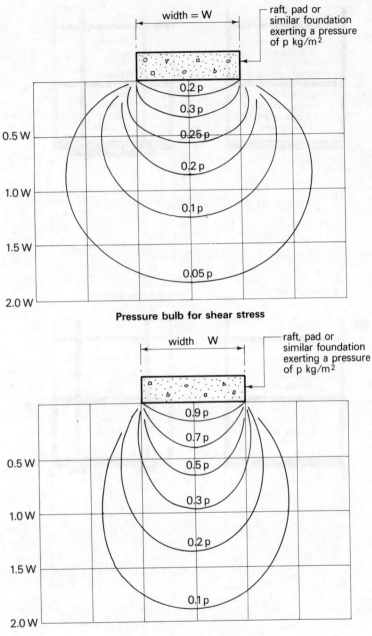

width = W

raft, pad or
similar foundation
exerting a pressure
of p kg/m²

0.2 p
0.3 p
0.5 W — 0.25 p
0.2 p
1.0 W — 0.1 p
1.5 W — 0.05 p
2.0 W —

Pressure bulb for shear stress

width   W

raft, pad or
similar foundation
exerting a pressure
of p kg/m²

0.9 p
0.7 p
0.5 W — 0.5 p
0.3 p
1.0 W — 0.2 p
1.5 W —
0.1 p
2.0 W —

Pressure bulb for vertical stress

Fig II.2 Raft or similar foundations — typical pressure bulbs

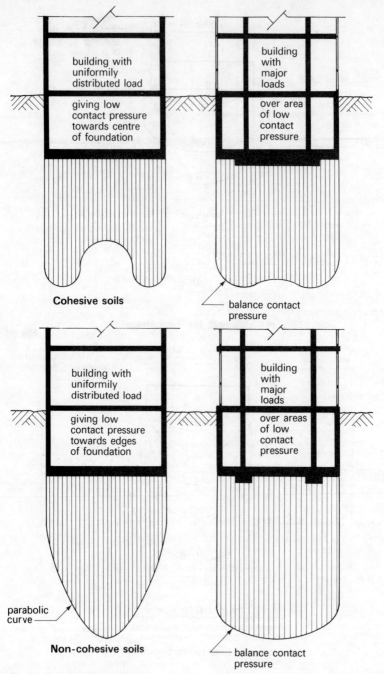

Cohesive soils

balance contact pressure

Non-cohesive soils

parabolic curve

balance contact pressure

Fig II.3 Typical contact pressures

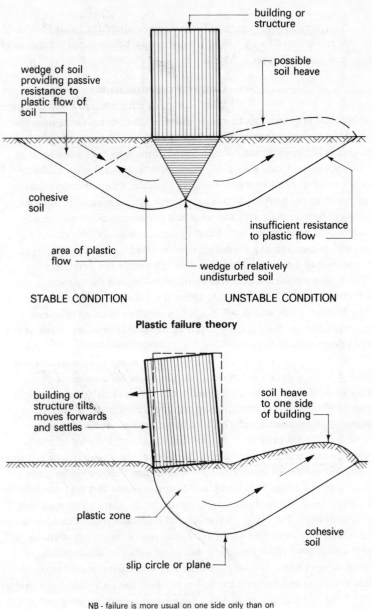

building or
structure

wedge of soil
providing passive
resistance to
plastic flow of
soil

possible
soil heave

cohesive
soil

insufficient resistance
to plastic flow

area of plastic
flow

wedge of relatively
undisturbed soil

STABLE CONDITION

UNSTABLE CONDITION

Plastic failure theory

building or
structure tilts,
moves forwards
and settles

soil heave
to one side
of building

plastic zone

cohesive
soil

slip circle or plane

NB - failure is more usual on one side only than on
both sides of the building or structure
failure can occur if pressure applied is about
six times the shear stress of the soil

Fig II.4   Plastic failure of foundations

4. *Pile foundations* — method for structures where the loads have to be transmitted to a point at some distance below the general ground level — see Chapter 5, Volume 3.

Combined foundations

When designing a foundation the area of the base must be large enough to ensure that the load per unit area does not exceed the bearing capacity of the soil upon which it rests. The ideal situation when considering column foundations is to have a square base of the appropriate area with the column positioned centrally. Unfortunately this ideal condition is not always possible. When developing sites in urban areas the proposed structure is very often required to abut an existing perimeter wall with the proposed perimeter columns in close proximity to the existing wall. Such a situation would result in an eccentric column loading if conventional isolated bases were employed. One method of overcoming this problem is to place the perimeter columns on a reinforced concrete continuous column foundation in the form of a strip. The strip is designed as a beam with the columns acting as point loads which will result in a negative bending moment occurring between the columns requiring top tensile reinforcement in the strip.

If sufficient plan area cannot be achieved by using a continuous column foundation a combined column foundation could be considered. This form of foundation consists of a reinforced concrete slab placed at right angles to the line of columns linking together an outer or perimeter column to an inner column. The base slab must have sufficient area to ensure that the load per unit area will not cause overstressing of the soil leading to unacceptable settlement. To alleviate the possibility of eccentric loading the centres of gravity of the columns and slab should be designed to coincide — see Fig. II.5. Combined foundations of this type are suitable for a pair of equally loaded columns or where the outer column carries the lightest load. The effect of the column loadings on the slab foundation is similar to that described above for continuous column foundations where a negative bending moment will occur between the columns.

In situations where the length of the slab foundation is restricted or where the outer column carries the heavier load the slab plan shape can be in the form of a trapezium with the widest end located nearest to the heavier loaded column — see Fig. II.5. As with the rectangular base the trapezoidal base will have negative bending moments between the columns and to eliminate eccentric loading the centres of gravity of columns and slab should coincide.

Alternative column foundations to the combined and trapezoidal forms

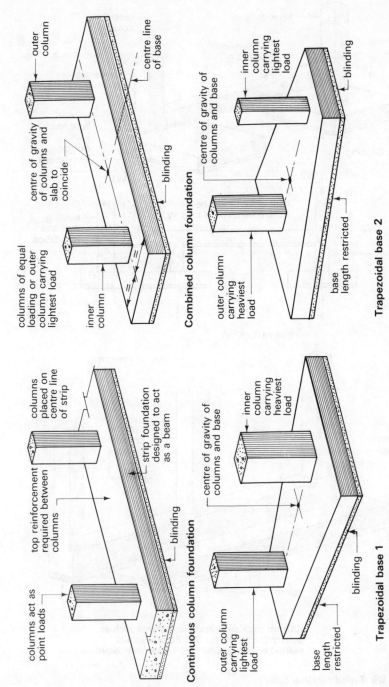

Continuous column foundation

columns placed on centre line of strip

columns act as point loads

top reinforcement required between columns

strip foundation designed to act as a beam

blinding

Combined column foundation

outer column

columns of equal loading or outer column carrying lightest load

inner column

centre of gravity of columns and slab to coincide

centre line of base

blinding

Trapezoidal base 1

outer column carrying lightest load

inner column carrying heaviest load

centre of gravity of columns and base

base length restricted

blinding

Trapezoidal base 2

inner column carrying lightest load

outer column carrying heaviest load

centre of gravity of columns and base

base length restricted

blinding

Fig II.5 Typical combined foundations

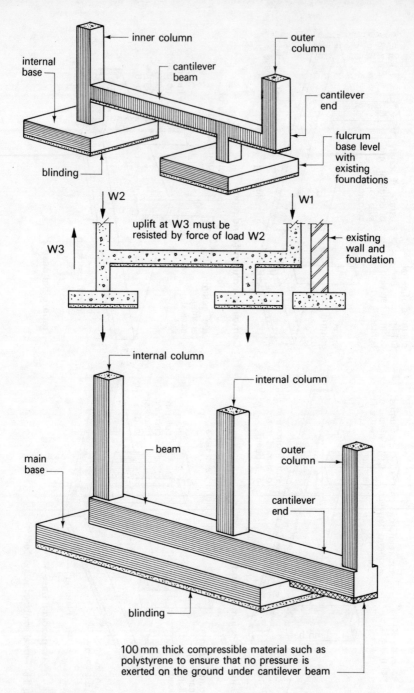

Fig II.6 Typical cantilever bases

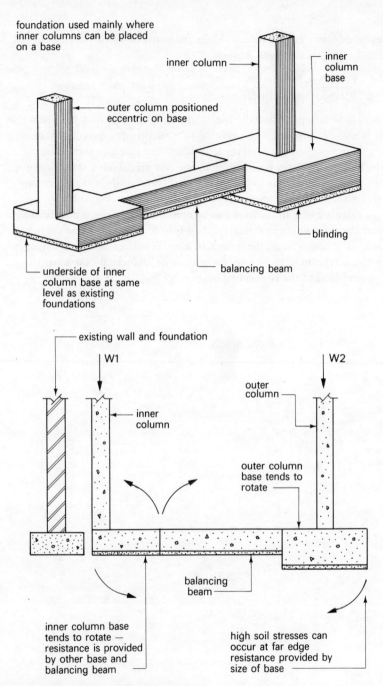

foundation used mainly where
inner columns can be placed
on a base

inner column

inner
column
base

outer column positioned
eccentric on base

blinding

balancing beam

underside of inner
column base at same
level as existing
foundations

existing wall and foundation

W1

W2

inner
column

outer
column

inner
column

outer column
base tends to
rotate

balancing
beam

inner column base
tends to rotate —
resistance is provided
by other base and
balancing beam

high soil stresses can
occur at far edge
resistance provided by
size of base

Fig II.7 Typical balanced base foundation

described above are the balanced base foundations which can be designed in one of two basic forms namely:

1. Cantilever foundations.
2. Balanced base foundations.

These foundations are usually specified where it is necessary to ensure that no pressure is imposed on an existing substructure or adjacent service such as a drain or sewer. A cantilever foundation can take two forms both of which use a cantilever beam which in one case acts about a short fulcrum column positioned near to the existing structure, while in the alternative version the beam cantilevers beyond a slab base — see Fig. II.6.

A balanced base foundation can be considered where it is possible to place the inner columns directly onto a base foundation. These columns are usually eccentric on the inner base causing within the base a tendency to rotate. This movement must be resisted or balanced by the inner column base and the connecting beam — see Fig. II.7.

8
Deep
basements

The construction of shallow single-storey basements and methods of waterproofing should have been covered in the previous three years of study and at this stage it is a useful exercise for students to recapitulate these basic concepts in terms of excavation, types, methods of construction, jointing techniques and waterproofing methods. These aspects are covered in the relevant chapters in Volumes 2 and 3.

One of the main concerns of the contractor and designer when constructing deep and/or multi-storey basements is the elimination as far as practicable of the need for temporary works such as timbering which in this context would be extensive, elaborate and costly. The solution is to be found in the use of diaphragm walling as the structural perimeter wall by constructing this ahead of the main excavation activity or alternatively by using a reinforced concrete land sunk open caisson if the subsoil conditions are favourable — see Fig. I.28.

Diaphragm walls can be designed to act as pure cantilevers in the first instance, but this is an expensive and usually unnecessary method. Therefore the major decision is one of providing temporary support until the floors which offer lateral restraint can be constructed or, alternatively, providing permanent support if the deep basement is to be constructed free of intermediate floors. Diaphragm walls can take the form of *in situ* reinforced concrete walling installed using the bentonite slurry trench system (see Fig. I.6, Volume 3), contiguous piling techniques as described in Chapter 5 or by using one of the precast concrete diaphragm walls.

PRECAST CONCRETE DIAPHRAGM WALLS

The main concept of precast concrete diaphragm walls is based on the *in situ* reinforced concrete walling installed using a bentonite slurry filled trench but has the advantages obtained by using factory produced components. The wall is constructed by forming a bentonite slurry filled trench of suitable width and depth and inserting into this the precast concrete panel or posts and panels according to the system being employed. As a normal bentonite slurry mix would not effectively seal the joints between the precast concrete components, a special mixture of bentonite and cement with a special retarder additive to control setting time is used. The precast units are placed into position within this mixture which will set sufficiently within a few days to enable excavation to take place within the basement area right up to the face of the diaphragm wall. To ensure that a clean wall face is exposed upon excavation it is usual practice to coat the proposed exposed wall faces with a special compound to reduce adhesion between the faces of the precast concrete units and the slurry mix.

The usual formats for precast concrete diaphragm walls are either simple tongue and groove jointed panels or a series of vertical posts with precast concrete infill panels as shown in Fig. II.8. If the subsoil is suitable it is possible to use a combination of precast concrete and *in situ* reinforced concrete to form a diaphragm wall by installing precast concrete vertical posts or beams at suitable centres and linking these together with an *in situ* reinforced concrete wall constructed in 2.000 deep stages as shown in Fig. II.9. The main advantage of this composite walling is the greater flexibility in design.

CONSTRUCTION METHODS

Four basic methods can be used in the construction of deep basements without the need for timbering or excessive timbering and these can be briefly described thus:

1. A series of lattice beams are installed so that they span between the tops of opposite diaphragm walls enabling the walls to act initially as propped cantilevers receiving their final lateral restraint when the internal floors have been constructed − see Fig. II.10.
2. The diaphragm walls are exposed by carrying out the excavation in stages and ground anchors are used to stabilise the walls as the work proceeds. This is a very popular method where no lateral restraint in the form of floors is to be provided − see Fig. II.10. The technique and details of ground anchor installations are given in conjunction with prestressing systems in Chapter 10.

92

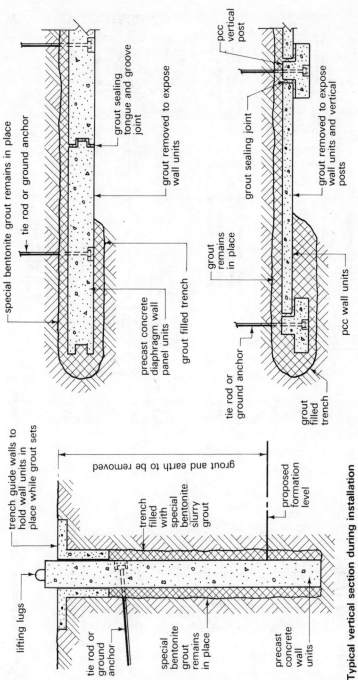

Typical vertical section during installation

Fig II.8 Typical precast concrete panel diaphragm wall details

special bentonite grout remains in place

tie rod or ground anchor

grout sealing tongue and groove joint

grout removed to expose wall units

precast concrete diaphragm wall panel units

grout filled trench

pcc vertical post

grout sealing joint

grout removed to expose wall units and vertical posts

grout remains in place

pcc wall units

tie rod or ground anchor

grout filled trench

trench guide walls to hold wall units in place while grout sets

grout and earth to be removed

trench filled with special bentonite slurry grout

proposed formation level

lifting lugs

tie rod or ground anchor

special bentonite grout remains in place

precast concrete wall units

93

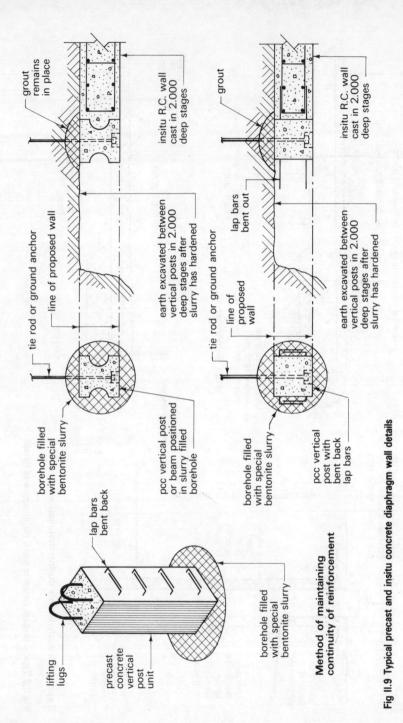

grout remains in place

tie rod or ground anchor

line of proposed wall

insitu R.C. wall cast in 2.000 deep stages

earth excavated between vertical posts in 2.000 deep stages after slurry has hardened

borehole filled with special bentonite slurry

pcc vertical post or beam positioned in slurry filled borehole

grout

tie rod or ground anchor

lap bars bent out

line of proposed wall

insitu R.C. wall cast in 2.000 deep stages

earth excavated between vertical posts in 2.000 deep stages after slurry has hardened

borehole filled with special bentonite slurry

pcc vertical post with bent back lap bars

lifting lugs

lap bars bent back

precast concrete vertical post unit

borehole filled with special bentonite slurry

Method of maintaining continuity of reinforcement

Fig II.9 Typical precast and insitu concrete diaphragm wall details

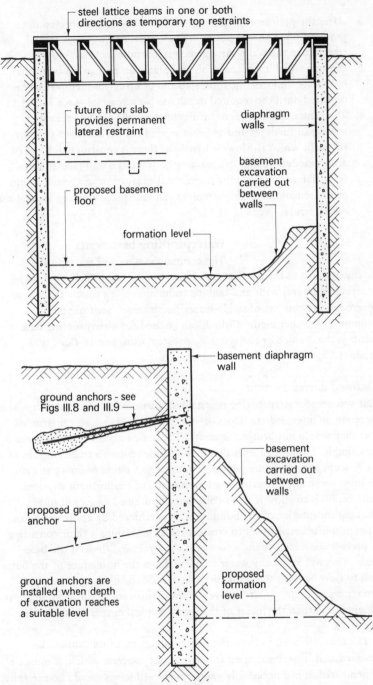

steel lattice beams in one or both directions as temporary top restraints

future floor slab provides permanent lateral restraint

diaphragm walls

proposed basement floor

basement excavation carried out between walls

formation level

basement diaphragm wall

ground anchors - see Figs III.8 and III.9

basement excavation carried out between walls

proposed ground anchor

ground anchors are installed when depth of excavation reaches a suitable level

proposed formation level

Fig II.10 Deep basement construction methods 1

95

3. After the perimeter diaphragm walls have been constructed the ground floor slab and beams are cast providing top edge lateral restraint to the walls. An opening is left in this initial floor slab through which men, materials and plant can pass to excavate the next stage and cast the floor slab and beams. This method can be repeated until the required depth has been reached — see Fig. II.11.

4. The centre area between the diaphragm walls can be excavated leaving an earth mound or berm around the perimeter to support the walls whilst the lowest basement floor is constructed. Slots to accommodate raking struts acting between the wall face and the floor slab are cut into the berm; final excavation and construction of the remainder of the basement can take place working around the raking struts — see Fig. II.11.

Waterproofing basements

The common methods of waterproofing basements such as the application of membranes, using dense monolithic concrete structural walls and tanking techniques using mastic asphalt, are covered in the context of single-storey basements — see Chapter 6, Volume 2. Another method which can be used for waterproofing basements is the drained or cavity tanking system using special floor tiles produced by the Atlas Stone Company Ltd.

Drained cavity system

This system of waterproofing is suitable for basements of any depth where there are no intermediate floors or where such floors are so constructed that they would not bridge the cavity. The basic concept of this system is very simple in that it accepts that it is possible to have a small amount of water seepage or moisture penetration through a dense concrete structural perimeter wall and therefore the best method of dealing with any such penetration is to allow it to be collected and drained away without it reaching the interior of the building. This is achieved by constructing an inner non-loadbearing wall to create a cavity and laying a floor consisting of precast concrete Dryangle tiles over the structural floor of the basement. This will allow any water trickling down the inside face of the outer wall to flow beneath the floor tiles where it can be discharged into the surface water drains or alternatively pumped into the drains if these are at a higher level than the invert of the sump. Typical details are shown in Fig. II.12.

The concrete used to form the structural basement wall should be designed as if it were specified for waterproof concrete which, it must be remembered, is not necessarily vapourproof. All joints should be carefully

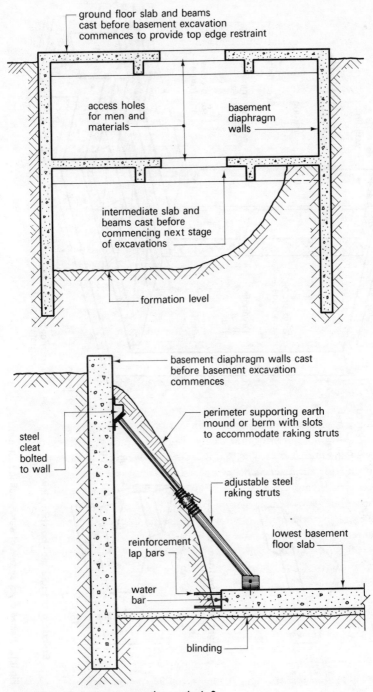

ground floor slab and beams cast before basement excavation commences to provide top edge restraint

access holes for men and materials

basement diaphragm walls

intermediate slab and beams cast before commencing next stage of excavations

formation level

basement diaphragm walls cast before basement excavation commences

perimeter supporting earth mound or berm with slots to accommodate raking struts

steel cleat bolted to wall

adjustable steel raking struts

lowest basement floor slab

reinforcement lap bars

water bar

blinding

Fig II.11 Deep basement construction methods 2

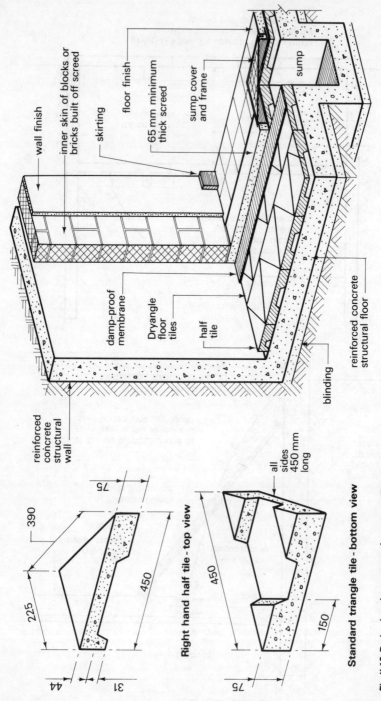

wall finish

inner skin of blocks or bricks built off screed

skirting

floor finish

65 mm minimum thick screed

sump cover and frame

sump

damp-proof membrane

Dryangle floor tiles

half tile

reinforced concrete structural floor

blinding

reinforced coincrete structural wall

390

75

225

450

44

31

Right hand half tile - top view

all sides 450 mm long

450

150

450

75

Standard triangle tile - bottom view

Fig II.12 Drained cavity system of waterproofing basements

designed, detailed and constructed including the fixing of suitable water bars — see Fig. II.6, Volume 2. It must be emphasised that the drained cavity system is designed to deal only with seepage which may occur and therefore the use of a dense concrete perimeter wall is an essential component in the system. The dense concrete used in the outer wall needs to be well vibrated to obtain maximum consolidation and this can usually be obtained by using poker vibrators.

The *in situ* concrete used in any form of basement construction can be placed by using chutes, pumps or tremie pipes. The placing of *in situ* concrete using pumps is covered in Part VII dealing with contractors' plant but the placing of concrete by means of a tremie pipe is worthy of special consideration.

Tremie pipes

A tremie pipe is a means of placing *in situ* concrete below the ground or water level where segregation of the mix must be avoided and can be used in the construction of piled foundations, basements and diaphragm walls. For work below ground level a rigid tube of plastic or metal can be used or alternatively a flexible PVC tube can be employed, the latter often being referred to as elephant trunking. In all cases the discharge end of the tremie pipe is kept below the upper level of the concrete being placed and the upper end of the pipe is fitted with a feed hopper attachment to receive the charges of concrete. As the level of the placed concrete rises the tube is shortened by removing lengths of pipe as required. The tremie pipe and its attached hopper head must be supported as necessary throughout the entire concrete placing operation.

Placing concrete below water level by means of a tremie pipe, very often suspended from a crane, requires more care and skill than placing concrete below ground level. The pipe should be of adequate diameter for the aggregates being used, common tube diameters being 150 and 200 mm. It should be watertight and have a smooth bore. The operational procedure can be enumerated thus:

1. Tremie pipe is positioned over the area to be concreted and lowered until the discharge end rests on the formation level.
2. A rabbit or travelling plug of plastic or cement bags is inserted into the top of the pipe to act as a barrier between the water and concrete. The weight of the first charge concrete will force the plug out of the discharge end of the tube.
3. When filled with concrete the tremie pipe is raised so that the discharge end is just above the formation level to allow the plug to be completely displaced and thus enable the concrete to flow.

4. Further concrete charges are introduced into the pipe and allowed to discharge within the concrete mass already placed, the rate of flow being controlled by raising and lowering the tremie pipe.
5. As the depth of the placed concrete increases the pipe is shortened by removing tube sections as necessary.

When placing concrete below the water level by the tremie pipe method great care must be taken to ensure that the discharge end of the pipe is not withdrawn clear of the concrete being placed as this could lead to a weak concrete mix by water mixing with the concrete. One tremie pipe will cover an area of approximately 30 m^2 and if more than one tremie pipe is being used simultaneous placing is usually recommended.

Sulphate-bearing soils

A problem which can occur when using concrete below ground level is the deterioration of the set cement due to sulphate attack by naturally occurring sulphates such as calcium, magnesium and sodium sulphates which are found particularly in clay soils. The main factors which influence the degree of attack are:

1. Amount and nature of sulphate in the soil.
2. Level of water table and amount of ground water movement.
3. Form of construction.
4. Type and quality of concrete.

The permanent removal of the ground water in the vicinity of the concrete structure is one method of limiting sulphate attack but this is very often impracticable and/or uneconomic. The only real alternative is to use a fully compacted concrete of low permeability. Once the chemical reaction between the sulphates and the set cement has taken place further deterioration can only come about when fresh sulphates are brought into contact with the concrete by ground water movement. Therefore concrete which is situated above the water table is not likely to suffer to a great extent.

Generally, precast concrete components suffer less than their *in situ* counterparts owing to the better control possible during casting under factory conditions. Basement walls and similar structural members usually have to resist lateral water pressure on one side only which increases the risk of water penetration and if evaporation can occur from the inner face sulphate attack can take place throughout the wall thickness.

Ordinary Portland cement manufactured to the recommendations of BS 12 usually contains a significant amount of tricalcium aluminate which is reacted upon by sulphates in the soil resulting in an expansion of the set cement causing a breakdown of the concrete. Sulphate-resisting Portland cement manufactured to the recommendations of BS 4027 has the proportion of tricalcium aluminate present limited to a maximum of 3.5% which will give the set cement a considerable resistance to sulphate attack. The minimum cement content and maximum free water : cement ratio of a concrete to be placed in a sulphate-bearing soil is related to the concentration of sulphate present in the soil and students seeking further information are advised to consult the table given in the Building Research Establishment Digest No. 250.

Part III
Prestressed
concrete

9
Principles and applications

The basic principle of prestressing concrete is very simple. If a material has little tensile strength it will fracture immediately its own tensile strength is exceeded, but if such a material is given an initial compression then when load-creating tension is applied the material will be able to withstand the force of this load as long as the initial compression is not exceeded. At this stage in the study of construction technology students will already be familiar with the properties of concrete that result in a material of high compressive strength with low tensile strength and that by inserting into the member steel reinforcing bars of the correct area and fixed to a predetermined pattern ordinary concrete can be given an acceptable amount of tensile strength. Prestressing techniques are applied to concrete in an endeavour to make full use of the material's high compressive strength.

Attempts were made at the end of the nineteenth century to induce a prestress into concrete but these were largely unsuccessful since the prestress could not be maintained. A French civil engineer Marie Eugene Leon Freyssinet (1879–1962) showed in the early 1920s how this problem could be overcome and demonstrated the type of concrete and prestressing steel which was required. His most significant contributions were the quantitative assessment of creep, shrinkage and the realisation that only a high strength steel at a high stress would achieve a permanent prestress in concrete.

In normal reinforced concrete the designer is unable to make full use of the high tensile strength of steel or of the high compressive strength of the concrete. When loaded above a certain limit tension cracks will occur in a reinforced concrete member and these should not generally be greater than 0.3 mm in width as recommended in BS 8110. This stage of cracking will normally be reached before the full strength potential of both steel and concrete has been obtained. In prestressed concrete the steel is stretched and securely anchored; it will then try to regain its original length but since it is fully restricted it will be subjecting the concrete to a compressive force throughout its life. A comparison of methods is shown in Fig. III.1.

Concrete whilst curing will shrink; it will also suffer losses in cross-section due to creep when subjected to pressure. Shrinkage and creep in concrete can normally be reduced to an acceptable level by using a material of high strength with a low workability. Mild steel will also suffer from relaxation losses which is the phenomenon of the stresses in steel under load decreasing towards a minimum value after a period of time. This can be counteracted by increasing the initial stress in the steel. If mild steel is used to induce a compressive force into a concrete member the amount of shrinkage, creep and relaxation which occurs will cancel out any induced stress. The special alloy steels used in prestressing, however, have different properties enabling the designer to induce extra stress into the concrete member, thus counteracting any losses due to shrinkage and creep and at the same time maintaining the induced compressive stress in the concrete component.

The high quality strength concrete specified for prestress work should take into account the method of stressing. For pre-tensioned work a minimum 28-day cube strength of 40 N/mm^2 is required whereas for post-tensioned work a minimum 28-day cube strength of 30 N/mm^2 is required. Steel in the form of wire or bars used for prestressing should conform to the recommendations of BS 5896 which covers steel wire manufactured from cold drawn plain carbon steel. The wire can be plain round, crimped or indented with a diameter range of 2 to 7 mm. Crimped and indented bars will develop a greater bond strength than plain round bars and are available in 4, 5 and 7 mm diameters. Another form of stressing wire or tendon is strand which consists of a straight core wire around which is helically wound further wires to form a 6 over 1 or 7 wire strand or a 9 over 9 over 1 giving a 19 wire strand tendon.

Seven wire strand is the easiest to manufacture and is in general use for tendon diameters up to 15 mm. The wire used to form the strand is cold drawn from plain carbon steel as recommended in BS 5896. To ensure close contact of the individual wires in the tendon the straight core wire is usually 2% larger in diameter than the outer wires which are helically wound around it at a pitch of 12 to 16 times the nominal diameter of the strand. Nineteen wire strands with diameters ranging from 25 to 32 mm

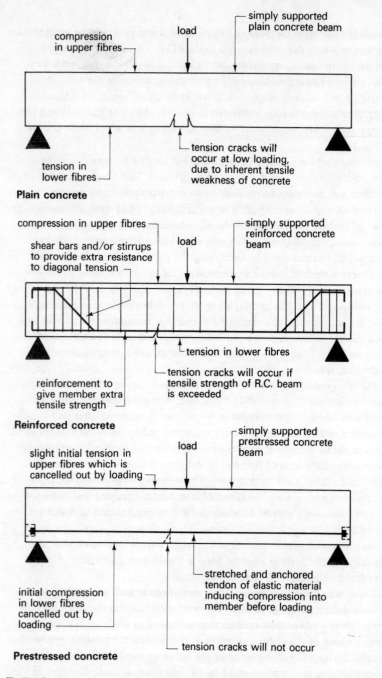

Fig III.1 Structural concrete — comparison of methods

are made from wires cold drawn from patented plain carbon steel to the recommendations of BS 4757. The straight core wire is covered with a helically wound layer of nine small diameter wires which are in turn covered with a helically wound layer of nine larger diameter wires. The helical pitch is similar to that given above for seven wire strand and the use of varying diameters is again to ensure close contact of the individual wires and to create good flexibility in the finished strand.

Tendons of strand can be used singly or in groups to form a multi-strand cable. The two major advantages of using strand are:

1. A large prestressing force can be provided in a restricted area.
2. It can be produced in long flexible lengths and can therefore be stored on drums thus saving site space and reducing site labour requirements by eliminating the site fabrication activity.

Having now considered the material requirements the basic principles of prestressing can be considered. A prestressing force inducing precompression into a concrete member can be achieved by anchoring a suitable tendon at one end of the member and applying an extension force at the other end which can be anchored when the desired extension has been reached. Upon release the anchored tendon in trying to regain its original length will induce a compressive force into the member. Figure III.2 shows a typical arrangement in which the tendon inducing the compressive force is acting about the neutral axis and is stressed so that it will cancel out the tension induced by the imposed load W. The stress diagrams show that the combined or final stress will result in a compressive stress in the upper fibres equal to twice that of the imposed load. The final stress must not exceed the characteristic strength of the concrete as recommended in BS 8110 and if the arrangement given in Fig. III.2 is adopted the stress induced by the imposed load will only be half its maximum.

To obtain a better economic balance the arrangement shown in Fig. III.3 is normally adopted where the stressing tendon is placed within the lower third of the section. The basic aim is to select a stress that when combined with the dead load will result in a compressive stress in the lower fibres equal to the maximum stresses induced by any live loads, resulting in a final stress diagram having in the upper fibres a compressive stress equal to the characteristic strength of the concrete and a zero stress in the bottom fibres. It must be observed that this is the pure theoretical case and is almost impossible to achieve in practice, but provided any induced tension occurring in the lower fibres is not in excess of the tensile strength of the concrete used, an acceptable prestressed condition will exist.

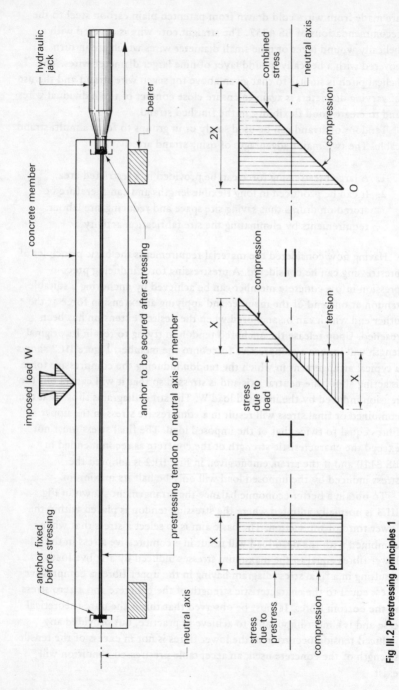

Fig III.2 Prestressing principles 1

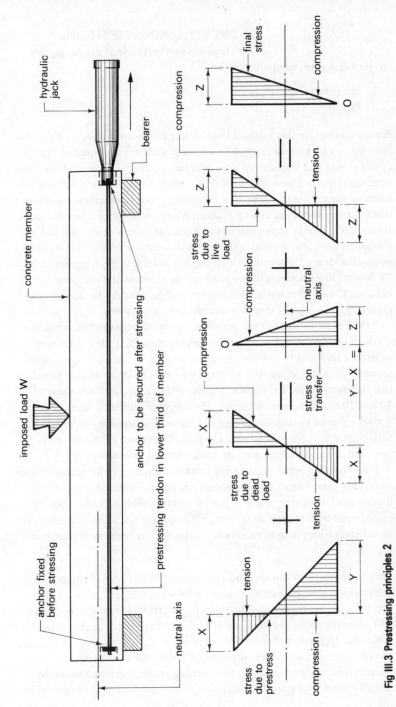

Fig III.3 Prestressing principles 2

PRESTRESSING METHODS

There are two methods of producing pre-stressed concrete members namely:

1. Pre-tensioning.
2. Post-tensioning.

Pre-tensioning: in this method the wires or cables are stressed before the concrete is cast around them. The stressing wires are anchored at one end of the mould and stressed by hydraulic jacks from the other end until the required stress has been obtained. It is common practice to overstress the wires by some 10% to counteract the anticipated losses which will occur due to creep, shrinkage and relaxation. After stressing the wires the side forms of the mould are positioned and the concrete is placed around the tensioned wires; the casting is then usually steam cured for 24 hours to obtain the desired characteristic strength, a common specification being 28 N/mm^2 in 24 hours. The wires are cut or released and the bond between the stressed wires and concrete will prevent the tendons from regaining their original length thus inducing the prestress.

At the extreme ends of the members the bond between the stressed wires and concrete will not be fully developed owing to low frictional resistance resulting in a contraction and swelling at the ends of the wires forming what is in effect a cone shape anchor. The distance over which this contraction takes place is called the transfer length and is equal to 80 to 120 times the wire diameter. Usually small diameter wires (2 to 5 mm) are used so that for any given total area of stressing wire a greater surface contact area is obtained. The bond between the stressed wires and concrete can also be improved by using crimped or indented wires.

Pre-tensioning is the prestressing method used mainly by manufacturers of precast components such as floor units and slabs employing the long line method of casting where precision metal moulds up to 120.000 long can be used with spacers or dividing plates positioned along the length to create the various lengths required — a typical arrangement is shown in Fig. III.4.

Post-tensioning: in this method the concrete is cast around ducts in which the stressing tendons can be housed and the stressing is carried out after the concrete has hardened. The tendons are stressed from one or both ends and when the stress required has been reached the tendons are anchored at their ends to prevent them returning to their original length thus inducing the compressive force. The anchors used form part of the finished component. The ducts for housing the stressing tendons can be formed by using flexible steel tubing or inflatable rubber tubes. The void created by

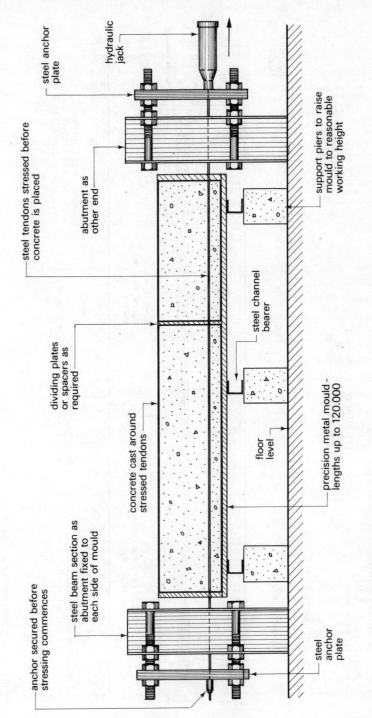

steel anchor plate

hydraulic jack

steel tendons stressed before concrete is placed

abutment as other end

support piers to raise mould to reasonable working height

dividing plates or spacers as required

steel channel bearer

concrete cast around stressed tendons

floor level

precision metal mould – lengths up to 120.000

anchor secured before stressing commences

steel beam section as abutment fixed to each side of mould

steel anchor plate

Fig III.4 Typical pre-tensioning arrangement

109

110

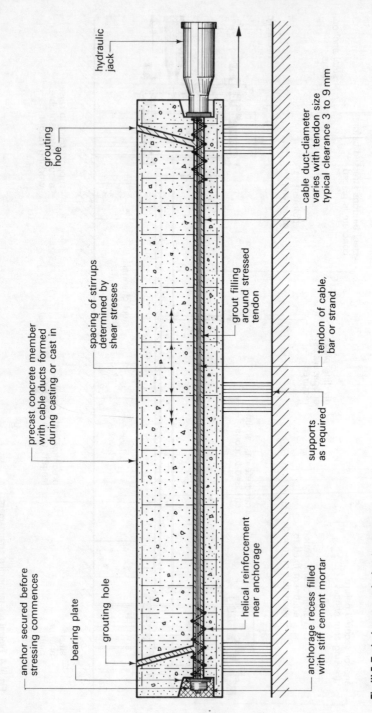

hydraulic jack

grouting hole

cable duct-diameter varies with tendon size typical clearance 3 to 9 mm

precast concrete member with cable ducts formed during casting or cast in

spacing of stirrups determined by shear stresses

grout filling around stressed tendon

tendon of cable, bar or strand

supports as required

anchor secured before stressing commences

bearing plate

grouting hole

helical reinforcement near anchorage

anchorage recess filled with stiff cement mortar

Fig III.5 Typical post-tensioning arrangement

the ducting will enable the stressing cables to be threaded prior to placing the concrete, or they can be positioned after the casting and curing of the concrete has been completed. In both cases the remaining space within the duct should be filled with grout to stop any moisture present setting up a corrosive action and to assist in stress distribution. A typical arrangement is shown in Fig. III.5.

Post-tensioning is the method usually employed where stressing is to be carried out on site, curved tendons are required, the complete member is to be formed by joining together a series of precast concrete units and where negative bending moments are encountered. Figure III.6 shows diagrammatically various methods of overcoming negative bending moments at fixed ends and for continuous spans. Figure III.7 shows a typical example of the use of curved tendons in the cross members of a girder bridge. Another application of post-tensioning is in the installation of ground anchors.

GROUND ANCHORS

A ground anchor is basically a prestressing tendon embedded and anchored into the soil to provide resistance to structural movement of a member by acting on a 'tying back' principle. Common applications include anchoring or tying back retaining walls and anchoring diaphragm walls particularly in the context of deep excavations. This latter application also has the advantage of providing a working area entirely free of timbering members such as struts and braces. Ground anchors can also be used in basement and similar constructions for anchoring the foundation slab to resist uplift pressures and to prevent floatation especially during the early stages of construction.

Ground anchors are known by their method of installation such as grouted anchors or by the nature of subsoil into which they are embedded such as rock anchors.

Rock anchors — these have been used successfully for many years and can be formed by inserting a prestressing bar into a predrilled hole. The leading end of the bar has expanding sleeves which grip the inside of the bored hole when the bar is rotated to a recommended torque to obtain the desired grip. The anchor bar is usually grouted over the fixed or anchorage length before being stressed and anchored at the external face.

Alternatively the anchorage of the leading end can be provided by grout injection relying on the bond developed between a ribbed sleeve and the wall of the bored hole. A dense high strength grout is required over the fixed length to develop sufficient resistance to pull out when the tendon is stressed. The unbonded or elastic length will need protection against

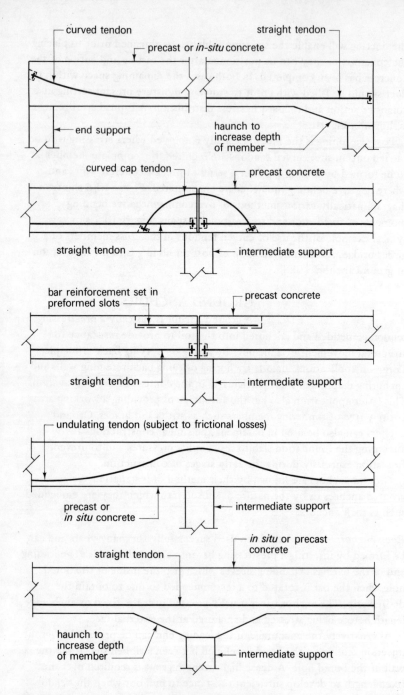

Fig III.6 Prestressing — overcoming negative bending moments

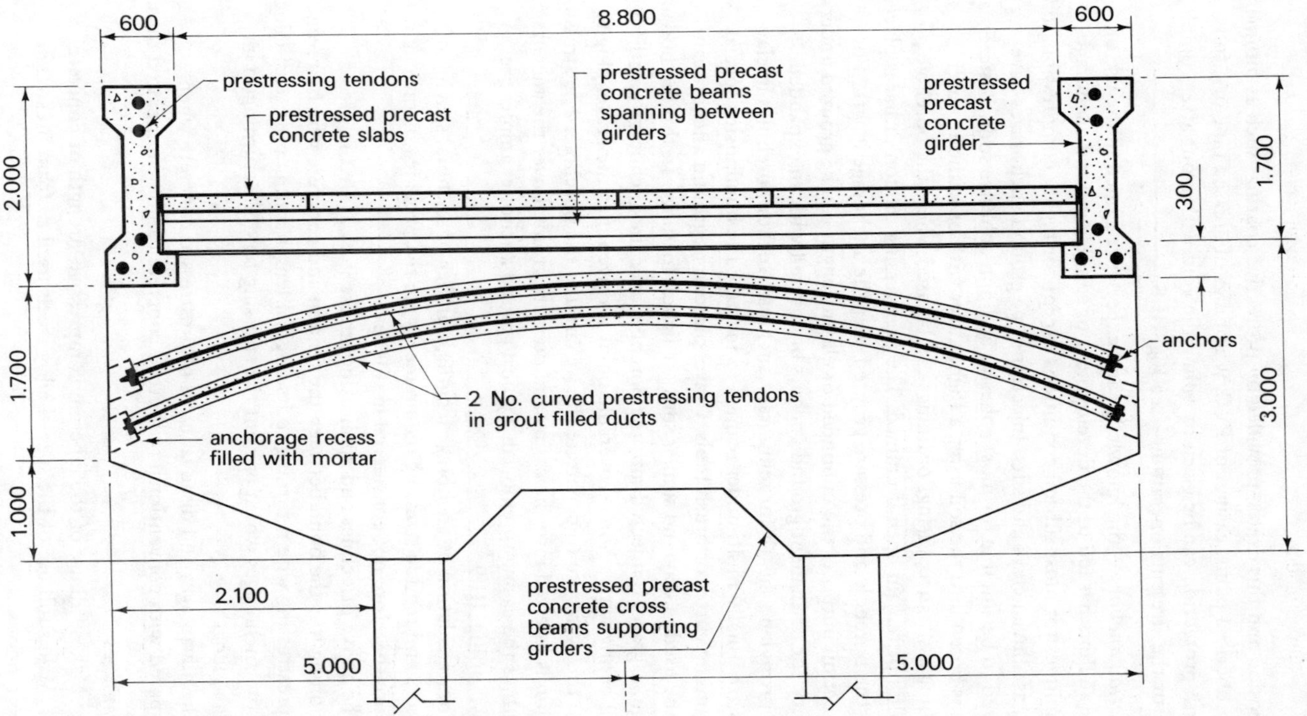

600
8.800
600
prestressing tendons
prestressed precast concrete slabs
prestressed precast concrete beams spanning between girders
prestressed precast concrete girder
2.000
1.700
300
anchors
1.700
2 No. curved prestressing tendons in grout filled ducts
anchorage recess filled with mortar
3.000
1.000
2.100
prestressed precast concrete cross beams supporting girders
5.000
5.000

Fig III.7 Typical example of the structural use of prestressed concrete

corrosion and this can be provided by protective coatings such as bitumen or rubberised paint, casings of PVC, wrappings of greased tape or a full-length protection can be given by filling the void with grout after completion of the stressing operation — see Fig. III.8.

Injection anchors — the knowledge and experience gained in the use of rock anchors has led to the development of suitable ground anchorage techniques for most subsoil conditions except for highly compressible soils such as alluvial clays and silts. Injection-type ground anchorages have proved to be suitable for most cohesive and non-cohesive soils. Basically a hole is bored into the soil using a flight auger with or without water flushing assistance; casings or linings can be used where the borehole would not remain open if unlined. The prestressing tendon or bar is placed into the borehole and pressure grouted over the anchorage length. For protection purposes the unbonded or elastic length can be grouted under gravity for permanent ground anchors or covered with an expanded polypropylene sheath for temporary anchors. Anchor boreholes in clay soils are usually multi-under-reamed to increase the bond, using special expanding cutter or brush tools. Gravel placement ground anchors can also be used in clay and similar soils for lighter loadings. In this method an irregular gravel is injected into the borehole over the anchorage length. A small casing with a non-recoverable point is driven into the gravel plug to force the aggregate to penetrate the soil around the borehole. The stressing tendon is inserted into the casing and pressure grouted over the anchorage length as the casing is removed. Typical ground anchor examples are shown in Fig. III.9.

The calculation of anchorage lengths, number of anchors, spacing of anchors and tendon stressing requirements are the province of the engineer and are therefore not considered in this text.

The above principles and applications make it clear that in pretensioning it is the bond between the tendon and concrete which prevents the prestressing wire returning to its original length and in post-tensioning it is the anchorages which prevent the stressing tendon returning to its original length.

The advantages and disadvantages of prestressed concrete when compared with conventional reinforced concrete can be enumerated thus:

Advantages
1. Makes full use of the inherent compressive strength of concrete.
2. Makes full use of the special alloy steels used to form prestressing tendons.
3. Eliminates tension cracks thus reducing the risk of corrosion of steel components.

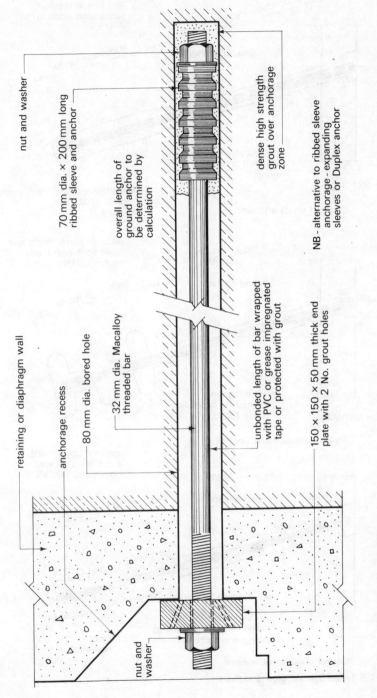

nut and washer

70 mm dia. × 200 mm long ribbed sleeve and anchor

overall length of ground anchor to be determined by calculation

dense high strength grout over anchorage zone

NB - alternative to ribbed sleeve anchorage - expanding sleeves or Duplex anchor

retaining or diaphragm wall

anchorage recess

80 mm dia. bored hole

32 mm dia. Macalloy threaded bar

unbonded length of bar wrapped with PVC or grease impregnated tape or protected with grout

150 × 150 × 50 mm thick end plate with 2 No. grout holes

nut and washer

Fig III.8 Typical rock ground anchor

115

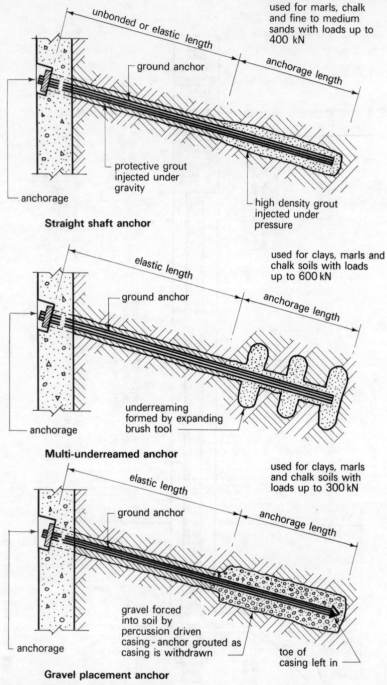

used for marls, chalk and fine to medium sands with loads up to 400 kN

unbonded or elastic length

ground anchor

anchorage length

protective grout injected under gravity

high density grout injected under pressure

anchorage

Straight shaft anchor

used for clays, marls and chalk soils with loads up to 600 kN

elastic length

ground anchor

anchorage length

underreaming formed by expanding brush tool

anchorage

Multi-underreamed anchor

used for clays, marls and chalk soils with loads up to 300 kN

elastic length

ground anchor

anchorage length

gravel forced into soil by percussion - driven casing - anchor grouted as casing is withdrawn

toe of casing left in

anchorage

Gravel placement anchor

Fig III.9 Typical injection ground anchors

116

4. Reduction in shear stresses.
5. For any given span and loading condition a member with a smaller cross-section can be used giving a reduction in weight.
6. Individual units can be joined together to act as a single member.

Disadvantages
1. High degree of control of materials, design and workmanship is required.
2. Special alloy steels are dearer than mild steels.
3. Extra cost of special equipment required to carry out stressing activities.

As a general comparison between the two structural mediums under consideration it is usually found that:

1. Up to 6.000 span traditional reinforced concrete is the most economic method.
2. Spans between 6.000 and 9.000 the two mediums are compatible.
3. Over 9.000 span prestressed concrete is generally more economical than reinforced concrete.

10

Prestressed concrete systems

The prestressing of concrete is usually carried out by a specialist contractor or alternatively by the main contractor using a particular system and equipment. The basic conception and principles of prestressing are common to all systems; it is only the type of tendon, type of anchorage and stressing equipment which varies. The following systems are typical and representative of the methods available.

BBRV (Simon-Carves Ltd)

This is a unique system of prestressing developed by four Swiss engineers namely Birkenmaier, Brandestini, Ros and Vogt whose initials are used to name the system. It differs from other systems in that multi-wire cables capable of providing prestressing forces from 300 to 7 800 kN are used and each wire is anchored at each end by means of enlarged heads formed on the wire. The cables or tendons are purpose made to suit individual requirements and may comprise up to 121 wires. The high tensile steel wire conforming to the recommendations of BS 5896 is cut to the correct length, any sheathing required is threaded on together with the correct anchorage before the button heads are formed using special equipment. The completed tendons can be coiled or left straight for delivery to site. Four types of stressing anchor are available, the choice being dependent on the prestressing force being induced, whereas the fixed anchors come in three forms. The finished tendon is fixed in the correct position to the formwork before concreting; when the concrete has hardened the tendons are stressed and grout is injected into the sheathing. If unsheathed tendons are to be drawn into preformed ducts the anchor is omitted from one and is fixed after drawing through the tendon, the button heads being formed with a portable machine.

Tendons in this system are tensioned by using a special hydraulic jack of the centre hole type with capacities ranging from 30 to 800 tonnes. A pull rod or pull sleeve depending on anchor type is coupled to the basic element carrying the wires. A lock nut, stressing stool, hydraulic jack and dynamometer are then threaded on. The applied stressing force can be read off the dynamometer whereas the actual extension achieved can be seen by the scale engraved on the jack.

After stressing the lock nut is tightened up and the jack released before grouting takes place. Losses due to friction, shrinkage and creep can be overcome by restressing at any time after the initial stressing operation so long as the tendons have not been grouted.

In common with most other prestressing systems tendons in long continuous members can be stressed in stages without breaking the continuity of the design. The first tendon length is stressed and grouted before the second tendon length is connected to it using a coupling anchor, after which it is stressed and grouted before repeating the procedure for any subsequent lengths. Alternatively, lengths of unstressed tendons can be coupled together and the complete tendon stressed from one or both ends. Figure III.10 shows a typical BBRV stressing arrangement.

CCL (CCL Systems Ltd)

This post-tensioning system uses a number of strands to form the tendon ranging from 4 to 31 strands according to the system being employed giving a range of prestressing forces from 450 to 5 000 kN. Two basic systems are available namely Cabco and Multiforce. In the Cabco system each strand or wire is stressed individually, the choice of hydraulic jack being governed by the size of strand being used. The system is fast and being manually operated eliminates the need for lifting equipment. By applying the total tendon force in stages problems such as differential elastic shortening and out of balance forces are reduced. Curved tendons are possible using this system without the need for spacers but spacers are recommended for tendons over 30.000 long.

The alternative Multiforce system uses the technique of simultaneously stressing all the strands forming the tendon. In both systems the basic anchorages are similar in design. The fixed or dead-end anchorage has a tube unit to distribute the load, a bearing plate and a compressing grip fitted to each strand. If the fixed anchorage is to be totally embedded in concrete a compressible gasket together with a bolted-on retaining plate are also used, a grout vent pipe being inserted into the grout hole of the tube unit. The compression grips consist of an outer sleeve with a

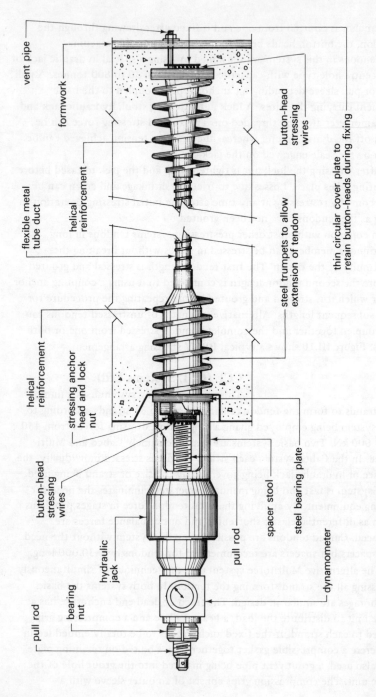

vent pipe

formwork

button-head
stressing
wires

flexible metal
tube duct

helical
reinforcement

steel trumpets to allow
extension of tendon

thin circular cover plate to
retain button-heads during fixing

helical
reinforcement

stressing anchor
head and lock
nut

button-head
stressing
wires

spacer stool

steel bearing plate

hydraulic
jack

dynamometer

pull rod

pull rod

bearing
nut

Fig III.10 Typical BBRV prestressing arrangement

machined and hardened insert which can be fitted whilst making up the tendon or installed after positioning the tendon. The stressing anchor is a similar device consisting of a tube unit, bearing plate and wedges working on a collet principle to secure the strands. The wedges are driven home to a 'set' by the hydraulic jack used for stressing the tendon before the jack is released and the stress is transferred. Figure III.11 shows a typical CCL Cabco stressing arrangement.

Dywidag (Dividag Stressed Concrete Ltd)

This post-tensioning system uses single or multiple bar tendons with diameters ranging from 12 to 36 mm for single bar applications and 16 mm diameter threaded bars for multiple bar tendons giving prestressing forces up to 950 kN for single bar tendons and up to 2 000 kN for multiple bar tendons. Both forms of tendon are placed inside a thin wall corrugated sheathing which is filled with grout after completion of the stressing operation. The single bar tendon can be of a smooth bar with cold rolled threads at each end to provide the connection for the anchorages, or alternatively it can be a threadbar which has a coarse thread along its entire length providing full mechanical bond. Threadbar is used for all multiple bar tendons.

Two forms of anchorage are available, namely the bell anchor and the square or rectangular plate anchor. During stressing, using a hydraulic jack acting against the bell or plate anchor, the tendon is stretched and at the same time the anchor nut is being continuously screwed down to provide the transfer of stress when the specified stress has been reached and the jack is released. A counter shows the number of revolutions of the anchor nut and the amount of elongation of the tendon. Like other similar systems the tendons can be restressed at any time before grouting. Figure III.12 shows typical details of a single bar tendon arrangement.

Macalloy (British Steel Corporation)

Like the method previously described the Macalloy system uses single or multiple bar tendons. The bars used are of a cold worked high strength alloy steel threaded at each end to provide an anchorage connection and are available in lengths up to 18.000 with nominal diameters ranging from 20 to 40 mm giving prestressing forces up to 875 kN for single bar tendons and up to 3500 kN using four 40 mm diameter bars to form the tendon.

The fixed anchorage consists of an end plate drilled and tapped to receive the tendon whereas the stressing anchor consists of a similar plate complete with a grouting flange and anchor nut. Stressing is carried out

121

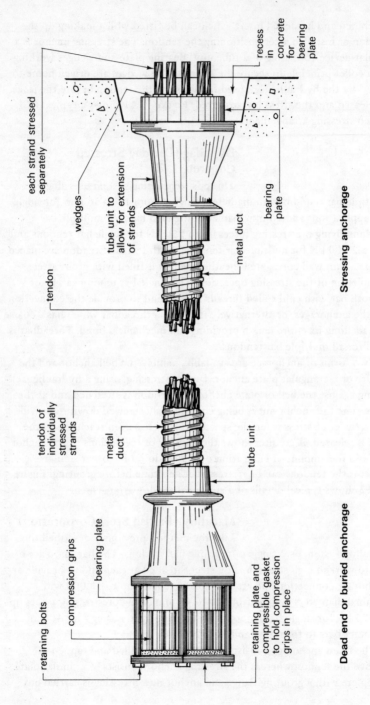

each strand stressed
separately

recess in
concrete
for bearing
plate

wedges

tube unit to
allow for extension
of strands

metal duct

tendon

bearing
plate

Stressing anchorage

tendon of
individually
stressed
strands

metal
duct

tube unit

retaining bolts

compression grips

bearing plate

retaining plate and
compressible gasket
to hold compression
grips in place

Dead end or buried anchorage

Fig III.11 Typical details of the CCL Cabco system of prestressing

122

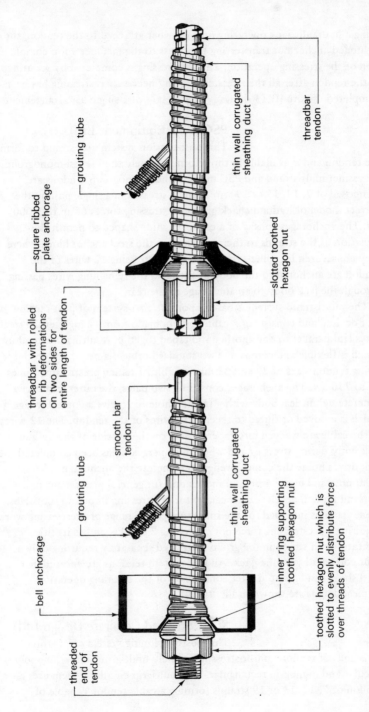

Fig III.12 Typical stressing anchors for Dywidag post-tensioning system

threadbar tendon

thin wall corrugated sheathing duct

grouting tube

square ribbed plate anchorage

slotted toothed hexagon nut

threadbar with rolled on rib deformations on two sides for entire length of tendon

smooth bar tendon

grouting tube

thin wall corrugated sheathing duct

bell anchorage

inner ring supporting toothed hexagon nut

toothed hexagon nut which is slotted to evenly distribute force over threads of tendon

threaded end of tendon

using a hydraulic jack operating on a drawbar attached to the tendon, the tightened anchor nut transferring the stress to the member upon completion of the stressing operation. The prestressing is completed by grouting in the tendon after all the stressing and any necessary restressing has been completed. Figure III.13 shows typical details of a single bar arrangement.

PSC (PSC Equipment Ltd)

This post-tension system uses strand to form the tendon and is available in three forms, namely the Freyssi-monogroup, Freyssinet multistrand and PSC monostrand. Monogroup tendons are composed of 7, 13, 15 or 19 wire strands stressed in a single pull by the correct model of hydraulic jack giving prestressing forces of up to 5 000 kN. The anchorages consist of a cast iron guide shaped to permit the deviation of the strands to their positions in the steel anchor block where they are secured by collet-type jaws. During stressing 12 wires of the tendon are anchored to the body of the jack, the remaining wires passing through the jack body to an anchorage at the rear.

The multistrand system is based upon the first system of post-tension ever devised and consists of a cable tendon made from 12 high-tensile steel wires laid parallel to one another and taped together resulting in a tendon which is flexible and compact. Two standard cable diameters are produced giving tendon sizes of 29 and 33 mm capable of taking prestressing forces up to 750 kN. The anchorages consist of two parts, the outer reinforced concrete cylindrical body with a tapered hole to receive a conical wedge which is grooved or fluted to receive the wires of the tendon. The 12 wires of the cable are wedged into tapered slots on the outside of the hydraulic jack body during stressing and when this operation has been completed the jack drives home the conical wedge to complete the anchorage.

Monostrand uses a 4 or 7 strand tendon for general prestressing or a 3 strand tendon developed especially for prestressing floor and roof slabs. This system is intended for the small to medium range of prestressing work requiring a prestressing force not exceeding 2 000 kN and as its title indicates each strand in the group is stressed separately requiring only a light compact hydraulic jack. All PSC system tendons are encased in a steel sheath and grouted after completion of the stressing operation. Typical details are shown in Fig. III.14.

SCD (Stressed Concrete Design Ltd)

This post-tensioning system offers three variations of tendon and/or stressing, namely multigrip circular, monogrip circular and monogrip rectangular. The multigrip circular system uses a tendon of 7, 12, 13 or 19 strands forming a cable tendon capable of

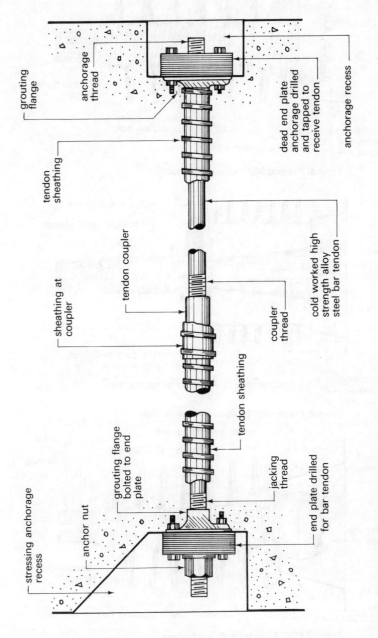

Fig III.13 Typical Macalloy prestressing system details

125

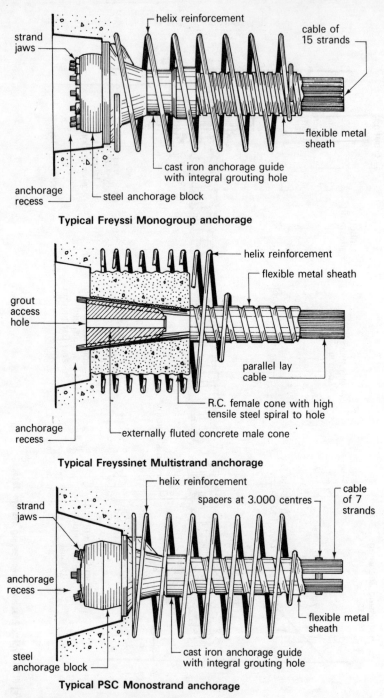

Typical Freyssi Monogroup anchorage

Typical Freyssinet Multistrand anchorage

Typical PSC Monostrand anchorage

Fig III.14 Typical PSC prestressing system

126

NB. in all cases anchorage zone helix reinforcement would be used according to design

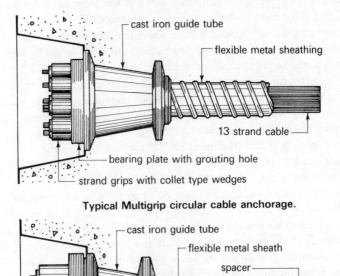

Typical **Multigrip** circular cable anchorage.

Typical **Monogrip** circular cable anchorage

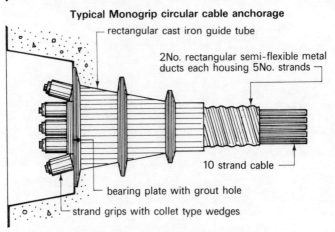

Typical **Rectangular** cable anchorage

Fig III.15 Typical SCD prestressing system details

accepting prestressing forces up to 5 000 kN. The anchorages consist of a cast iron guide plate enabling the individual strands of the cable to fan out and pass through a bearing plate where they are secured with steel collet-type wedges. The strands pass through the body of the hydraulic jack to a rear anchorage and are therefore stressed simultaneously.

The monogrip circular tendons are available as a single, 4, 7 or 12 strand cable in which each strand is stressed individually and the whole tendon being capable of taking prestressing forces up to 3 200 kN. Each strand in the tendon is separated from adjacent strands by means of circular spacers at 2.000 centres or less if the tendon is curved. The anchorages are similar in principle to those already described for multi-grip tendons.

The monogrip rectangular system by virtue of using a rectangular tendon composed of 3 to 27 strands capable of taking a prestressing force of up to 3 900 kN affords maximum eccentricity with a wide range of tendon sizes. The anchorage guide plate and bearing plates work on the same principle as described above for the multigrip method. In all methods the tendon is encased in a sheath unless preformed ducts have been cast; upon completion of the stressing operation all tendons are grouted in. See Fig. III.15 for typical details.

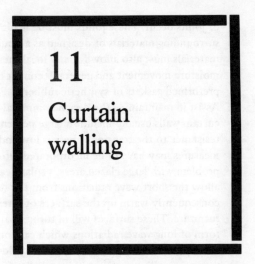

11
Curtain walling

Part IV
Claddings

Curtain walls are a form of external lightweight cladding attached to a framed structure forming a complete envelope or sheath around the structural frame. They are non-load-bearing claddings which have to support only their own deadweight and any imposed wind loadings which are transferred to the structural frame through connectors which are usually positioned at floor levels. The basic conception of most curtain walls is a series of vertical mullions spanning from floor to floor interconnected by horizontal transoms forming openings into which can be fixed panels of glass or infill panels of opaque materials. Most curtain walls are constructed by using a patent or proprietary system produced by metal window manufacturers.

The primary objectives of using curtain walling systems are:

1. Provide an enclosure to the structure which will give the necessary protection against the elements.
2. Make use of dry construction methods.
3. Impose onto the structural frame the minimum load in the form of claddings.
4. Exploit an architectural feature.

To fulfil its primary functions a curtain wall must meet the following requirements:

1. Resistance to the elements — the materials used in curtain walls are usually impervious and therefore in themselves present no problem but by virtue of the way in which they are fabricated a large number

129

of joints occur. These joints must be made as impervious as the surrounding materials or designed as a drained joint. The jointing materials must also allow for any local thermal, structural or moisture movement and generally consist of mastics, sealants and/or preformed gaskets of synthetic rubber or PVC.

2. Assist in maintaining the designed internal temperatures — since curtain walls usually include a large percentage of glass the overall resistance to the transfer of heat is low and therefore preventive measures may have to be incorporated into the design. Another problem with large glazed areas is solar heat gain since glass will allow the short wave radiations from the sun to pass through and consequently warm up the surfaces of internal walls, equipment and furniture. These surfaces will in turn radiate this acquired heat in the form of long wave radiations which cannot pass back through the glazing thus creating an internal heat build-up. Louvres fixed within a curtain walling system will have little effect upon this heat build-up but they will reduce solar glare. A system of non-transparent external louvres will slightly reduce the heat gain by absorbing heat and radiating it back to the external air. The usual methods employed to solve the problem of internal heat gain are:

(a) Deep recessed windows which could be used in conjunction with external vertical fins.

(b) Balanced internal heating and ventilation systems.

(c) Use of special solar control glass such as reflective glasses which during manufacture are modified by depositing on the surface of the glass a metallic or dielectric reflective layer. The efficiency of this form of glazing can be increased if the class is tilted by $5°$ to $15°$ to increase the angle of incidence.

3. Adequate strength — although curtain walls are classified as non-load-bearing they must be able to carry their own weight and resist both positive and negative wind loadings. The magnitude of this latter loading will depend upon three basic factors:

(a) Height of building;

(b) Degree of exposure;

(c) Location of building.

The strength of curtain walling relies mainly upon the stiffness of the vertical component or mullion together with its anchorage or fixing to the structural frame. Glazing beads and the use of compressible materials also add to the resistance of possible wind damage of the glazed and infill panel areas by enabling these units to move independently of the curtain wall framing.

130

4. Provide required degree of fire resistance — this is probably one of the greatest restrictions encountered when using curtain walling techniques because of the large proportion of unprotected areas as defined in Approved Document B supporting Building Regulation B4 — external fire spread (see Part IV, Volume 3). By using suitable materials or combinations of materials the opaque infill panels can normally achieve the required fire resistance to enable them to be classified as protected areas.

5. Easy to assemble and fix — the principal member of a curtain walling system is usually the mullion which can be a solid or box section which is fixed to the structural frame at floor levels by means of adjustable anchorages or connectors. The infill framing and panels may be obtained as a series of individual components or as a single prefabricated unit. The main problems are ease of handling, amount of site assembly required and mode of access to the fixing position.

6. Provide required degree of sound insulation — sound originating from within the structure may be transmitted vertically through the curtain walling members. The chief source of this form of structure-borne sound is machinery and this may be reduced by isolating the offending machines by mounting them on resilient pads and/or using resilient connectors in the joints between mullion lengths. Airborne sound can be troublesome with curtain walling systems since the lightweight cladding has little mass to offer in the form of a sound barrier, the weakest point being the glazed areas. A reduction in the amount of sound transmitted can be achieved by:

   (a) Reducing the areas of glazing.
   (b) Using sealed windows of increased glass thickness.
   (c) Double glazing in the form of inner and outer panes of glass with an air space of 150 to 200 mm between them.

7. Provide for thermal and structural movements — since curtain walling is situated on an external face of the structure it will be more exposed than the structural frame and will therefore be subject to greater amounts of temperature change resulting in high thermal movement. The main frame may also be subjected to greater settlement than the claddings attached to its outer face. These differential movements mean that the curtain walling systems should be so designed, fabricated and fixed that the attached cladding can move independently of the structure. The usual methods of providing for this required movement are to have slotted bolt connections and, to allow for movement within the curtain walling itself, to have spigot connections and/or mastic sealed joints. Figures IV.1 and IV.2 show typical curtain walling examples to illustrate these principles.

131

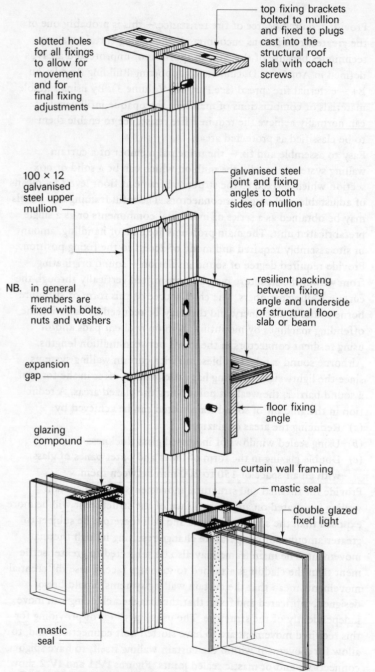

top fixing brackets bolted to mullion and fixed to plugs cast into the structural roof slab with coach screws

slotted holes for all fixings to allow for movement and for final fixing adjustments

100 × 12 galvanised steel upper mullion

galvanised steel joint and fixing angles to both sides of mullion

NB. in general members are fixed with bolts, nuts and washers

resilient packing between fixing angle and underside of structural floor slab or beam

expansion gap

floor fixing angle

glazing compound

curtain wall framing

mastic seal

double glazed fixed light

mastic seal

Fig IV.1 Typical curtain walling details 1

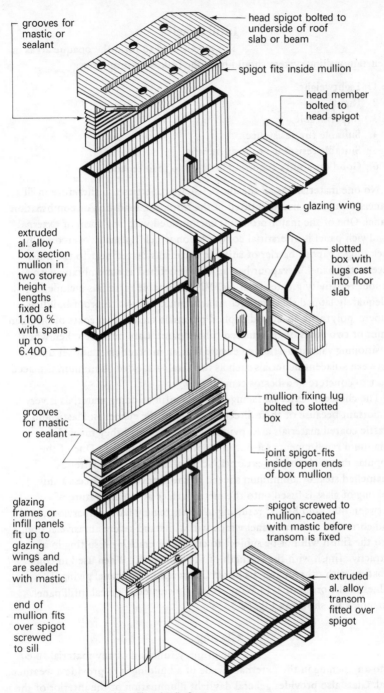

grooves for mastic or sealant

head spigot bolted to underside of roof slab or beam

spigot fits inside mullion

head member bolted to head spigot

glazing wing

slotted box with lugs cast into floor slab

extruded al. alloy box section mullion in two storey height lengths fixed at 1.100 % with spans up to 6.400

mullion fixing lug bolted to slotted box

grooves for mastic or sealant

joint spigot-fits inside open ends of box mullion

glazing frames or infill panels fit up to glazing wings and are sealed with mastic

spigot screwed to mullion-coated with mastic before transom is fixed

end of mullion fits over spigot screwed to sill

extruded al. alloy transom fitted over spigot

Fig IV.2 Typical curtain walling details 2

133

Infill Panels

The panels used to form the opaque areas in a curtain walling system should have the following properties:

1. Lightweight.
2. Rigid.
3. Impermeable.
4. Suitable fire resistance.
5. Suitable resistance to heat transfer.
6. Good durability requiring little or no maintenance.

No one material has all the above listed properties and therefore infill panels are usually manufactured in the form of a sandwich or combination panel. One of the major problems encountered with any form of external sandwich panel is interstitial condensation which is usually overcome by including a vapour barrier of suitable material situated near to the inner face of the panel. A vapour barrier can be defined as a membrane with a vapour resistivity greater than 100 MN/g. Suitable materials include adequately lapped sheeting such as aluminium foil, waterproof building papers, polythene sheet and applied materials such as two coats of bitumen paint or two coats of chlorinated rubber paint. Care must be taken when positioning vapour barriers to ensure that an interaction is not set up between adjacent materials such as the alkali attack of aluminium if placed next to concrete or asbestos cement.

The choice of external facing for these infill cladding materials is very important because of their direct exposure to the elements. Plastic and plastic coated materials are obvious choices provided they meet the minimum requirements set out in Approved Document B. One of the most popular materials for the external facing of infill panels is vitreous enamelled steel or aluminium sheets. In the preparation process a thin coating of glass is fused onto the metal surface at a temperature of between 800° and 860°C resulting in an extremely hard, impervious, acid and corrosion resistant panel which will withstand severe abrasive action, also the finish will not be subject to crazing or cracking resulting in an attractive finish with the strength of the base metal. When used in combination with other materials a thin lightweight infill panel giving 'U' values in the order of $1.14 \text{ W/m}^2 \text{ K}$ can be created. Typical infill panel examples are shown in Fig. IV.3.

Glazing

The primary function of any material fixed into an opening in the external façade of a building is to provide a weather seal. Glass also provides general daylight illumination of the interior of the

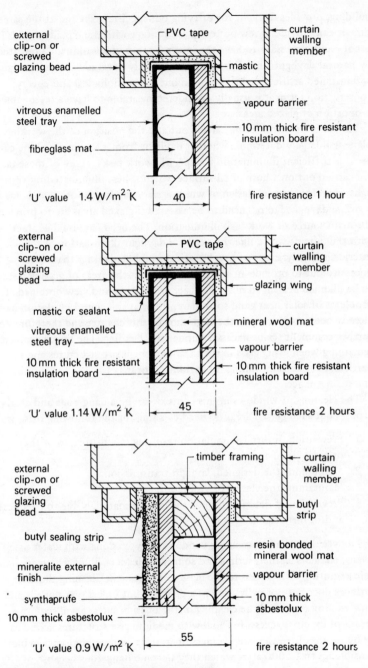

Fig IV.3 Typical curtain walling infill panel details

135

building, provides daylight for carrying out specific tasks and at the same time it can provide a view out or visual contact with the outside world. It is not really necessary to have the internal space of a building illuminated by natural daylight since this can be adequately covered by a well-designed and installed artificial lighting system but in psychological and energy conservation terms it is usually considered desirable to have a reasonable proportion of glazed areas.

The nature of work to be carried out and the position of the working plane will largely determine whether daylight from glazed areas alone can provide sufficient illumination for specific work tasks. Many of these tasks are carried out on a horizontal surface which is best illuminated by vertical light; therefore careful design of window size, window position and possible daylight factors need to be assessed if glazed areas are to provide the main source of work task illumination. The need for visual contact with the outside world like the need for daylight illumination of the interior is largely a case of psychological well-being rather than dire necessity but to provide an acceptable view out the areas of glazing need to be planned bearing in mind the size, orientation and view obtained. The problems of solar heat gain, solar glare, thermal and sound insulation have already been considered at the beginning of this chapter and therefore no further comment seems necessary under this heading. The major problem remaining when using glass in the façades of high rise structures and in particular for curtain walling is providing a means of access for cleaning and maintenance.

The cleaning of windows in this context can be a dangerous and costly process but such glazed areas do need cleaning for the following reasons:

1. Prevention of dirt accumulation resulting in a distortion of the original visual appearance.
2. Maintaining the designed daylight transmission.
3. Maintaining the clarity of vision out.
4. Prevention of deterioration of the glazing materials due to chemical and/or dirt attack.

The usual method of cleaning windows is by washing with water using swabs, chamois leather, scrims and squeegee cleaners, all of which are hand held requiring close access to the glass to be cleaned. Cleaning the internal surfaces does not normally present a problem but unless pivot or tilt and turn windows are used the cleaning of the outside surface will require a means of external access. In the low to medium rise structures access can be by means of trestles, step ladders or straight ladders; the latter being possible up to 11.000 after which they become dangerous because of flexing and the lack of overall control. Tower scaffolds are seldom used for window cleaning because of the cost and time involved in assembly but lightweight quickly erected scaffold systems could be considered for heights up to 6.000.

Access for external cleaning of curtain wall façades in high rise structures is generally by the use of suspended cradles which can be of a temporary nature as shown in Fig. VII.38, or alternatively they can be of a permanent system designed and constructed as an integral part of the structure. The simplest form is to install a universal beam section at roof level positioned about 450 mm in front of the general façade line and continuous around the perimeter of the roof. A conventional cradle is attached by means of castors located on the bottom flange of the ring or edge beam. Control is by means of ropes from the cradle which must be lowered to ground level for access purposes.

A better form is where the transversing track is concealed by placing a pair of rails to a 750 mm gauge to which is fixed a trolley having projecting beams or davits which can be retracted and/or luffed, the latter being particularly useful for negotiating projections in the general façade. The trolley may be manually operated from the cradle for transversing or it may be an electric-powered trolley giving vertical movement of between 5 and 15 m per minute or horizontal movement of 5 to 12 m per minute. This form of trolley is usually considered essential for heights over 45.000. In all cases the roof must have been structurally designed to accept the load of the apparatus.

If the height of suspension is over 30.000 it is a statutory requirement that some form of cable restraint is incorporated in the design to overcome the problem of unacceptable cradle movement due to the action of wind around the structure. Winds moving up the face of the building will cause the cradle to swing at right angles to the face giving rise to possible impact damage to the face of the structure as well as placing the operatives in a hazardous situation.

Crosswinds can cause the cradle to move horizontally along the face of the building with equally disastrous results. Methods of cradle and cable restraint available are the use of suction grips, eyebolts fixed to the façade through which the suspension ropes can be threaded, suspension ropes suitably tensioned at ground level, electrical cutouts at intervals of 15.000 which will prevent further cradle movement until a special plug through which the hoist line passes has been inserted and using the mullions as a guide for rollers which are either in contact with the mullion face to prevent lateral movement or by castors located behind a mullion flange to prevent outward movement.

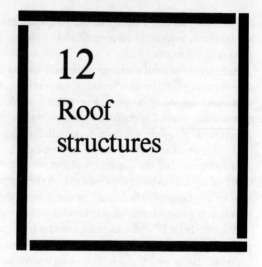

12
Roof structures

Part V
Roofs

Roof design and construction, like foundations, is an extensive topic and it is therefore the usual practice to build up a student's comprehension by including in each year of study one particular aspect of roofing techniques. In this text the roofs under consideration are those suitable for large clear spans using structural timber, steelwork and reinforced concrete. Other types of roof construction are described in earlier volumes.

LARGE SPAN TIMBER ROOFS

A wide variety of timber roofs are available for both medium and large spans and these can be classified under three headings:

1. Pitched trusses.
2. Flat top girders.
3. Bowstring trusses.

Pitched trusses: these are two-dimensional triangulated designed frames spaced at 4.500 to 6.000 centres with spans up to 30.000. The pitch should have a depth to span ratio of 1:5 or steeper and be chosen to suit roof coverings. The basic construction follows that of an ordinary domestic small span roof truss as shown in Fig. II.46, Volume 1 except that the arrangement and number of struts and ties will vary according to the type of truss being used — see Fig. V.1.

Flat top girders: basically these are lattice beams of low pitch spaced at

4.500 to 6.000 centres and can be economically used for spans up to 45.000 with a depth to span ratio of 1 : 8 to 1 : 10. Construction details are similar to those of the roof truss in that the joints and connections are usually made with timber connectors and bolts. The main advantage of this form of roof structure is the reduction in volume of the building which should result in savings in the heating installation required and in the running costs. Typical flat top girder outlines are shown in Fig. V.1 and typical construction details are shown in Fig. V.2.

Bowstring trusses: these trusses are basically a lattice girder with a curved upper chord and are spaced at 4.500 to 6.000 centres with an economic span range of up to 75.000. The depth-to-span ratio is usually 1 : 6 to 1 : 8 with the top chord radius approximately equal to the span. They can be constructed from solid segmental timber pieces but the chords are usually formed from laminated timber with solid timber struts and ties forming the lattice members — see Fig. V.3 for typical details. The older form of bowstring truss known as a belfast truss which has interlacing struts and ties is not very often specified because it is difficult to analyse fully the stresses involved and although relatively small section timbers can be used they are very expensive in labour costs.

Choice of truss and timber
To decide upon the most suitable and economic truss to be specified for any given situation the following should be considered:

1. Availability of suitable timber in the sizes required.
2. Cost of alternative timber.
3. Design and fabrication costs.
4. Transportation problems and costs.
5. On-site assembly and erection problems and costs.
6. Roof covering material availability and costs.
7. Architectural design considerations.

It may be possible that the consideration given to the last item may well outweigh some of the economic solutions found for the preceding items.

The timber specified should comply with the recommendations of BS 5268 and any subsequent amendments. The specified timber must also satisfy the structural stability requirements of Building Regulation A1 and the recommendations set out in the supportive Approved Document A. This means that unless a designer, manufacturer or builder makes special arrangements to obtain approval, each piece of timber used in a strength application in any form of building will have to be stress graded to the

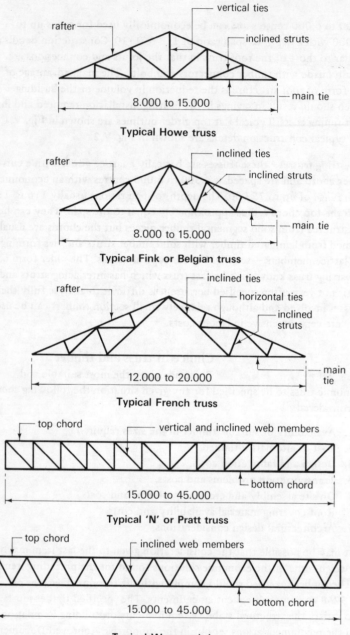

Fig V.1 Typical large span truss and girder types

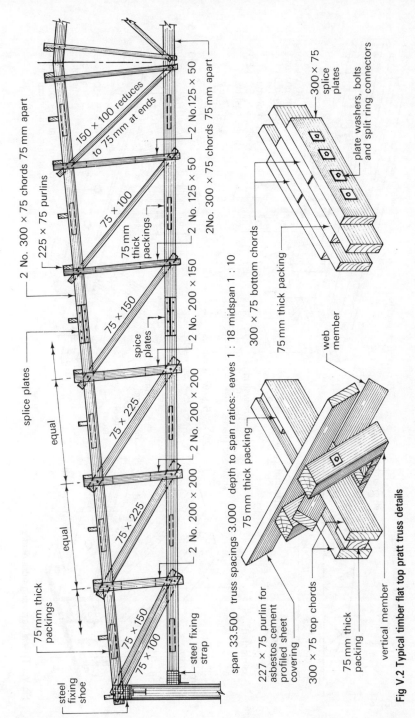

span 33.500 truss spacings 3.000 depth to span ratios:- eaves 1 : 18 midspan 1 : 10

Fig V.2 Typical timber flat top pratt truss details

141

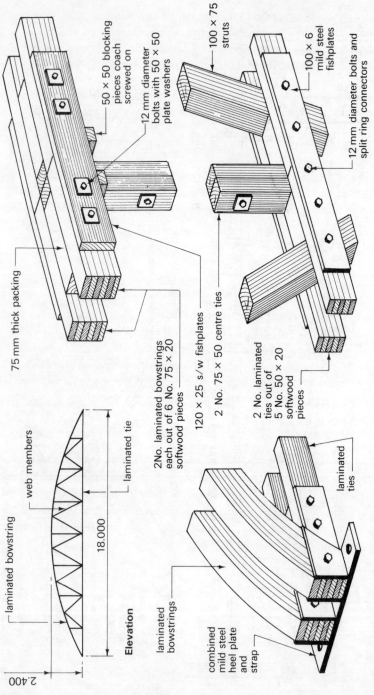

50 × 50 blocking pieces coach screwed on

100 × 75 struts

100 × 6 mild steel fishplates

12 mm diameter bolts with 50 × 50 plate washers

12 mm diameter bolts and split ring connectors

75 mm thick packing

2No. laminated bowstrings each out of 6 No. 75 × 20 softwood pieces

120 × 25 s/w fishplates

2 No. 75 × 50 centre ties

2 No. laminated ties out of 5 No. 50 × 20 softwood pieces

laminated ties

web members

laminated tie

laminated bowstring

18.000

2.400

Elevation

laminated bowstrings

combined mild steel heel plate and strap

Fig V.3 Typical bowstring truss details

142

grades set out in BS 5268. Reference should also be made to BS 4978 'Timber Grades for Structural Use'. This standard was published after collaboration between engineers, scientists and the timber trades of the United Kingdom and the main timber-exporting countries such as Canada, Finland and Sweden. The stress grading can be carried out either visually or by machine and the permissible stresses for the various grades and species are set out in BS 5268 : Part 2.

Visual stress grading is carried out by the knot area ratio (KAR) method in which the proportion of cross-section occupied by the projected area of knots is assessed. Any pieces of timber where the KAR is less than one-fifth is graded special structural (SS) and for pieces of timber where the KAR is between one-fifth and one-half is graded as general structural (GS) or SS depending on whether a margin condition exists. Any piece of timber with a KAR exceeding one-half is automatically rejected as being suitable for structural work.

Machine stress grading relies on the correlation between the strength of timber and its stiffness. These grading machines work on one of two principles, those which apply a fixed load and measure the deflection and those which measure the load required to cause a fixed deflection. The gradings obtained are designated MGS (machine general structural) or MSS (machine special structural) and these grades are comparable to the visual grades given above. BS 5268 also describes two further grades, namely M50 and M75 which means that these pieces of timber have been graded as having 50 or 75% of the strength of a clear specimen of a similar species.

All graded timber must be marked so that it can be immediately identified by the specifier, supplier and user. Visually graded timber is marked at least once within the length of each piece with GS or SS together with a mark to indicate the company or grader responsible for the grading. Machine stress graded timber should be marked MGS, MSS, M50 or M75 at least once within the length of each piece together with a mark to indicate species, grading machine used, BSI Kitemark and the relevant BS number, namely BS 4978. Machine stress graded timber can also be colour coded with a coloured dye at one end or a series of dashes throughout the length. The colour coding used being:

Green	MGS
Purple	MSS
Blue	M50
Red	M75

The most readily available species of structural softwoods are imported redwood, imported whitewood and imported commercial western hemlock. Other suitable structural softwoods are only generally available in

143

small quantities. Structural softwoods are supplied in standard lengths commencing at 1.800 and increasing by 300 mm increments to a maximum length of 6.300 with section size within the range of the table given in BS 4471. When specifying stress graded timber the following points should be considered:

1. Species.
2. Section size.
3. Length.
4. Preparation requirements.
5. Stress grade.
6. Moisture content.

Jointing

Connections between structural timber members may be made by:

1. Nails.
2. Screws.
3. Glue and nails or screws.
4. Truss plates.
5. Bolts.
6. Bolts and timber connectors.

Nails and screws are usually used in conjunction with plywood gussets and like the truss or gangnail plates are usually confined to the small to medium span trusses.

The usual fixings such as nails, screws and bolts have their own limitations. Cut nails will generally have a greater holding power than round wire nails due to the higher friction set up by their rough sharp edges and also the smaller disturbance of the timber grain. The joint efficiency of nails may be as low as 15% owing to the difficulty in driving sufficient numbers of them within a given area to obtain the required shear value. Screws have a greater holding power than nails but are dearer in both labour and material costs.

Joints made with a rigid bar such as a bolt usually have low efficiency due to the low shear strength of timber parallel to the grain and the unequal distribution of bearing stress along the shank of the bolt. The weakest point in these connections is where the high stresses are set up around the bolt and various methods have been devised to overcome this problem. The solution lies in the use of timber connectors which are designed to increase the bearing area of the bolts as described below.

Toothed plate timber connector — sometimes called bulldog connectors.

These are used to form an efficient joint without the need for special equipment and are suitable for all types of connections, especially when small sections are being used. To form the connection the timber members are held in position and drilled for the bolt to provide a bolt hole with 1.5 mm clearance. If the timber is not too dense the toothed connectors can be embedded by tightening up the permanent bolts. In dense timbers or where more than three connectors are used the embedding pressure is provided by a high tensile steel rod threaded at both ends. Once the connectors have been embedded the rod is removed and replaced by the permanent bolt.

Split ring timber connectors — these connectors are suitable for any type of structure, timbers of all densities, are very efficient and develop a high strength joint. The split ring is a circular band of steel with a split tongue and groove vertical joint. A special boring and cutting tool is required to form the bolt hole and the grooves in the face of the timber into which the connector is inserted making the ring independent of the bolt itself. The split in the ring ensures that a tight fit is achieved on the timber core but at the same time being sufficiently flexible to give a full bearing on the timber outside the ring when under heavy load.

Shear plate timber connector — these are counterparts of the split ring connectors, they are housed flush into the timber members and are used for demountable structures. See Fig. V.4 for typical timber connector details.

LARGE SPAN STEEL ROOFS

The roof types given for large span timber roofs can also be designed and fabricated using standard structural steel sections. Span ranges and the spacings of the frames or lattice girders are similar to those given for timber roofs. Connections can be of traditional gusset plates to which the struts and ties would be bolted or welded; alternatively an all-welded construction is possible especially if steel tubes are used to form the struts and ties. Large span steel roofs can also take the form of space decks and space frames.

Space decks

A space deck is a structural roofing system designed to give large clear spans with wide column spacings. It is based on a simple repetitive unit consisting of an inverted pyramid frame which can be joined to similar frames to give spans of up to 22.000 for single spanning designs and up to 33.000 for two-way spanning roofs. These basic

145

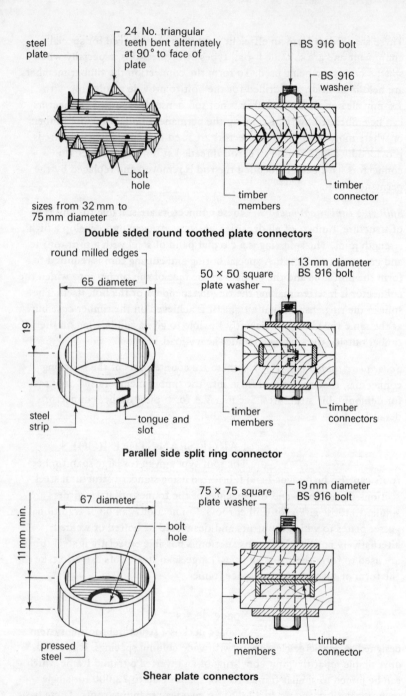

steel plate

24 No. triangular teeth bent alternately at 90° to face of plate

bolt hole

sizes from 32 mm to 75 mm diameter

BS 916 bolt

BS 916 washer

timber connector

timber members

Double sided round toothed plate connectors

round milled edges

65 diameter

19

steel strip

tongue and slot

50 × 50 square plate washer

13 mm diameter BS 916 bolt

timber members

timber connectors

Parallel side split ring connector

67 diameter

11 mm min.

bolt hole

pressed steel

75 × 75 square plate washer

19 mm diameter BS 916 bolt

timber members

timber connector

Shear plate connectors

Fig V.4 Typical BS 1579 timber connectors

146

units are joined together at the upper surface by bolting adjacent angle framing together and by fixing threaded tie bars between the apex couplers — see Fig. V.5.

Edge treatments include vertical fascia, mansard and cantilever. Roof-lights can be fixed directly over the 1.200 x 1.200 modular upper framing — see Fig. V.6. The roof can be laid or screeded to falls, or alternatively a camber to form the falls can be induced by tightening up the main tie bars. The space deck roof can be supported by steel or concrete columns or fixed to padstones or ring beams situated at the top of load-bearing perimeter walls.

The most usual and economic roof covering for space decks is an insulated decking of wood wool slabs 50 mm thick covered with three layers of built-up roofing felt with a layer of reflective chippings. The void created by the space deck structure can be used to house all forms of services and the underside can be left exposed or covered in with an attached or suspended ceiling. The units are usually supplied with a basic protective paint coating applied by a dipping process after the units have been degreased, shot blasted and phosphate treated to provide the necessary key.

The simplicity of the space deck unit format eliminates many of the handling and transportation problems encountered with other forms of large span roof. A complete roof can usually be transported on one lorry by stacking the pyramid units one inside the other. Site assembly and erection is usually carried out by a specialist sub-contractor who must have access to the whole floor area. Assembly is rapid with a small labour force which assembles the units as beams in the inverted position, turns them over and connects the whole structure together by adding the secondary tie bars. Two general methods of assembly and erection can be used:

1. Deck is assembled on the completed floor immediately below the final position and lifted directly into its final position.
2. Deck is assembled outside the perimeter of the building and lifted in small sections to be connected together in the final position — this is generally more expensive than method 1.

The main contractor has to provide prior to assembly and erection a clear, level and hard surface to the perimeter of the proposed building as well as to the whole floor area to be covered by the roof. These surfaces must be capable of accepting the load from a 25 tonne mobile crane. The main contractor is also responsible for unloading, checking, storing and protecting the units during and after erection and for providing all necessary temporary works and plant such as scaffolds, ladders and hoists. The site procedures and main contractor responsibilities set out above in

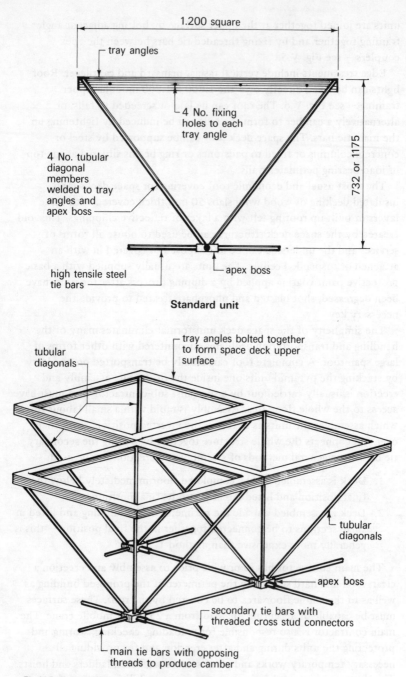

Fig V.5 Typical 'Space Deck' standard units

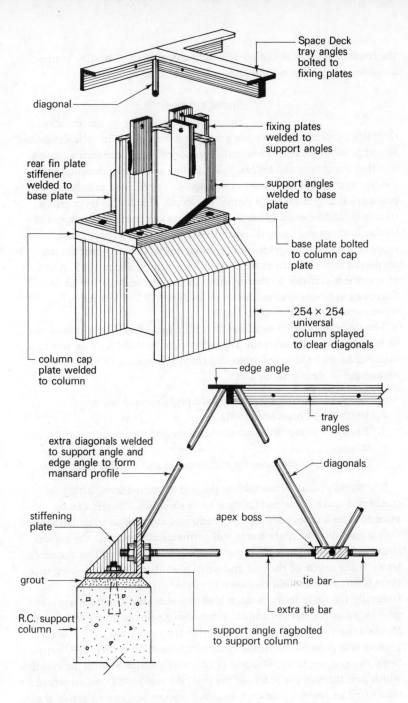

Space Deck
tray angles
bolted to
fixing plates

diagonal

fixing plates
welded to
support angles

rear fin plate
stiffener
welded to
base plate

support angles
welded to base
plate

base plate bolted
to column cap
plate

254 × 254
universal
column splayed
to clear diagonals

column cap
plate welded
to column

edge angle

tray
angles

extra diagonals welded
to support angle and
edge angle to form
mansard profile

diagonals

stiffening
plate

apex boss

grout

tie bar

R.C. support
column

extra tie bar

support angle ragbolted
to support column

Fig V.6 Typical 'Space Deck' edge fixing details

the context of space decks are generally common to all roofing contractors involving specialist sub-contractors and materials.

Space frames

Space frames are similar to space decks in their basic concept but they are generally more flexible in their design and layout possibilities since their main component is the connector joining together the chords and braces. Space frames are usually designed as a double-layer grid as opposed to the single-layer grid used mainly for geometrical shapes such as a dome. The depth of a double-layer grid is relatively shallow when compared with other structural roof systems of similar loadings and span, the span-to-depth ratio for a space frame supported on all its edges would be about 1:20 whereas a space frame supported near the corners would require a ratio of about 1:15. A variety of systems is available to the architect and builder and the British Steel Corporation Nodus system illustrated in Figs. V.7 and V.8 is a typical representative of such systems.

The claddings used in conjunction with a space frame roof should not be unduly heavy and normally any lightweight profiled decking would be suitable. As with the space decks described above some of the main advantages of these systems are:

1. Constructed from simple standard prefabricated units.
2. Units can be mass produced.
3. Roof can be rapidly assembled and erected on site using semi-skilled labour.
4. Small sizes of components make storage and transportation easy.

Site works consist of assembling the grid at ground level, lifting the completed space frame and fixing it to its supports. The grid can be assembled on a series of blocks to counteract any ground irregularities and during assembly the space frame will automatically generate the correct shape and camber. The correct procedure is to start assembling the space frame at the centre of the grid and work towards the edges ensuring that there is sufficient ground clearance to enable the camber to be formed. Generally the space frame is assembled as a pure roof structure but it is possible to install services and fix the cladding prior to lifting and fixing. Mobile cranes are usually employed to lift the completed roof structure holding it in position whilst the columns are erected and fixed. Alternatively the grid can be constructed in an offset position around the columns which pass through the spaces in the grid, the completed roof structure is then lifted and moved sideways onto the support seatings on top of the columns.

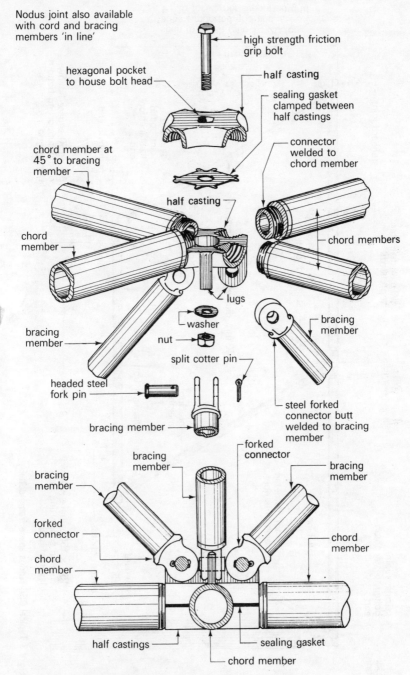

Nodus joint also available with cord and bracing members 'in line'

high strength friction grip bolt

hexagonal pocket to house bolt head

half casting

sealing gasket clamped between half castings

connector welded to chord member

chord member at 45° to bracing member

half casting

chord members

chord member

lugs

washer

nut

bracing member

bracing member

split cotter pin

headed steel fork pin

bracing member

steel forked connector butt welded to bracing member

forked connector

bracing member

bracing member

bracing member

bracing member

forked connector

chord member

chord member

chord member

half castings

sealing gasket

chord member

Fig V.7 Typical BSC Nodus System joint details.

151

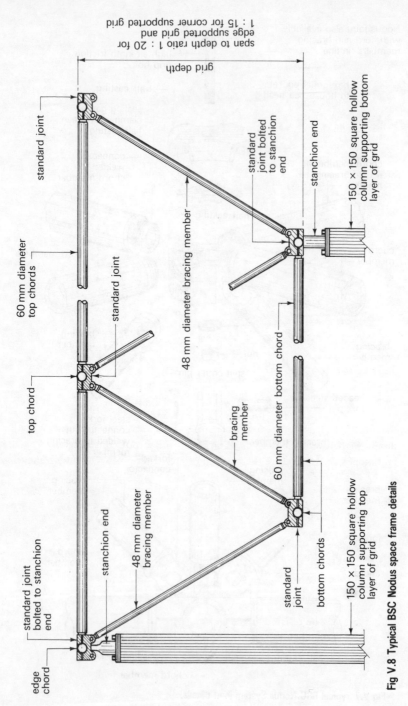

span to depth ratio 1 : 20 for
edge supported grid and
1 : 15 for corner supported grid

grid depth

standard joint

60 mm diameter top chords

standard joint

top chord

standard joint bolted to stanchion end

stanchion end

48 mm diameter bracing member

standard joint bolted to stanchion end

stanchion end

48 mm diameter bracing member

bracing member

60 mm diameter bottom chord

standard joint

bottom chords

150 × 150 square hollow column supporting top layer of grid

150 × 150 square hollow column supporting bottom layer of grid

edge chord

Fig V.8 Typical BSC Nodus space frame details

SHELL ROOFS

A shell roof may be defined as a structural curved skin covering a given plan shape and area, the main points being:

1. Primarily a structural element.
2. Basic strength of any particular shell is inherent in its shape.
3. Quantity of material required to cover a given plan shape and area is generally less than other forms of roofing.

The basic materials which can be used in the formation of a shell roof are concrete, timber and steel. Concrete shell roofs consist of a thin curved reinforced membrane cast *in situ* over timber formwork whereas timber shells are usually formed from carefully designed laminated timber and steel shells are generally formed using a single layer grid. Concrete shell roofs although popular are very often costly to construct since the formwork required is usually purpose made from timber and which is in itself a shell roof and has little chance of being re-used to enable the cost of the formwork to be apportioned over several contracts.

A wide variety of shell roof shapes and types can be designed and constructed but they can be classified under three headings:

1. Domes.
2. Vaults.
3. Saddle shapes and conoids.

Domes: in their simplest form these consist of a half sphere but domes based on the ellipse, parabola and hyperbola are also possible. Domes have been constructed by architects and builders over the centuries using individually shaped wedge blocks or traditional timber roof construction techniques. It is therefore the method of construction together with the materials employed rather than the geometrical setting out which has changed over the years.

Domes are double curvature shells which can be rotational and are formed by a curved line rotating around a vertical axis or they can be translational domes which are formed by a curved line moving over another curved line — see Figs. V.9 and 10. Pendentive domes are formed by inscribing within the base circle a polygon and cutting vertical planes through the true hemispherical dome.

Any dome shell roof will tend to flatten due to the loadings and this tendency must be resisted by stiffening beams or similar to all the cut edges. As a general guide domes which rise in excess of one-sixth of their diameter will require a ring beam. Timber domes like their steel counterparts are usually constructed on a single-layer grid system and covered with a suitable thin skin membrane.

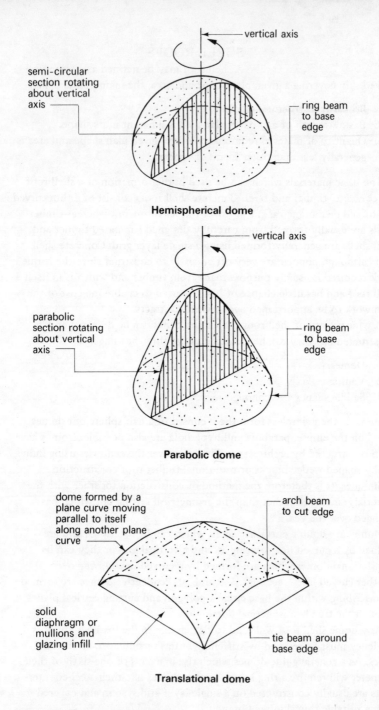

Fig V.9 Typical dome roof shapes 1

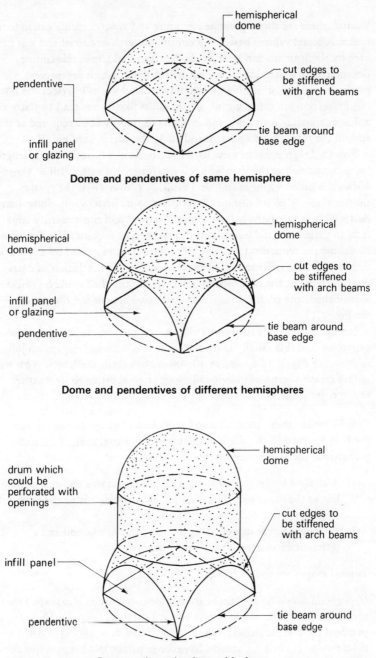

Dome and pendentives of same hemisphere

Dome and pendentives of different hemispheres

Dome and pendentives with drum

Fig V.10 Typical dome roof shapes 2

155

Vaults: these are shells of single curvature and are commonly called barrel vaults. A barrel vault is basically a continuous arch or tunnel and was first used by the Romans and later by the Norman builders in this country. Geometrically a barrel vault is a cut half cylinder which presents no particular setting out problems. When two barrel vaults intersect the lines of intersection are called groins. Barrel vaults like domes tend to flatten unless adequately restrained and in vaults restraint will be required at the ends in the form of a diaphragm and along the edges — see Fig. V.11.

From a design point of view barrel vaults act as a beam with the length being considered as the span which if it is longer than its width or chord distance is called a long span barrel vault, or conversely if the span is shorter than the chord distance is termed a short barrel vault. Short barrel vaults with their relatively large chord distances and consequently large radii to their inner and outer curved surfaces may require stiffening ribs to overcome the tendency to buckle. The extra stresses caused by the intro-duction of these stiffeners or ribs will necessitate the inclusion of extra reinforcement at the rib position, alternatively the shell could be thickened locally about the rib for distance of about one-fifth of the rib spacing — see Fig. V.11.

In large barrel vault shell roofs allowances must be made for thermal expansion and this usually takes the form of continuous expansion joints as shown in Fig. V.12 spaced at 30.000 centres along the length. This will in fact create a series of individually supported abutting roofs weather sealed together.

Conoid shells: these are similar to barrel vaults but are double curvature shells as opposed to the single curvature of the barrel vault. Two basic geometrical forms are encountered:

1. A straight line is moved along a curved line at one end and a straight line at the other end. The resultant shape being cut to the required length.
2. A straight line is moved along a curved line at one end and a different curved line at the other end.

Typical shapes are shown in Fig. V.13.

Hyperbolic paraboloids — these are double-curvature saddle-shaped shells formed geometrically by moving a vertical parabola over another vertical parabola set at right angles to the moving parabola — see Fig. V.14. The saddle shape created is termed a hyperbolic paraboloid because horizontal sections taken through the roof will give a hyperbolic outline and vertical sections will result in a parabolic outline. To obtain a more practical shape

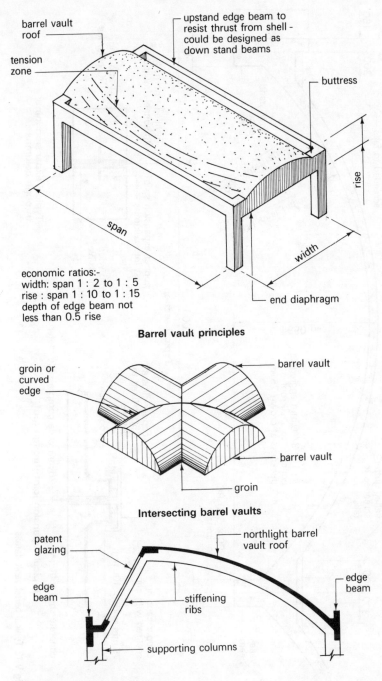

barrel vault roof

tension zone

upstand edge beam to resist thrust from shell - could be designed as down stand beams

buttress

rise

span

width

economic ratios:-
width: span 1 : 2 to 1 : 5
rise : span 1 : 10 to 1 : 15
depth of edge beam not less than 0.5 rise

end diaphragm

Barrel vault principles

groin or curved edge

barrel vault

barrel vault

groin

Intersecting barrel vaults

patent glazing

northlight barrel vault roof

edge beam

stiffening ribs

edge beam

supporting columns

Fig V.11 Typical barrel vaults

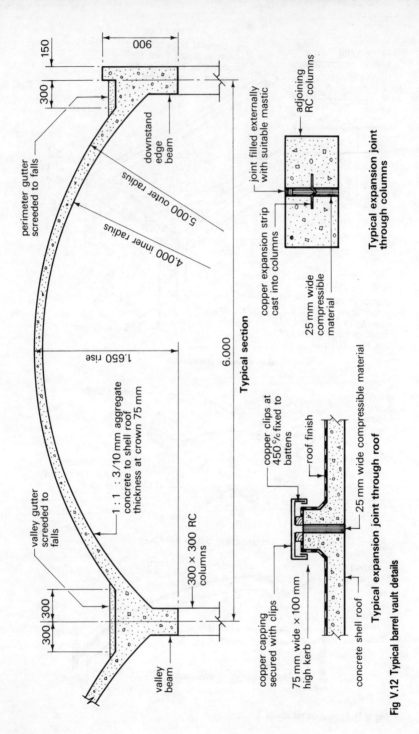

Typical section

900

150

300

perimeter gutter screeded to falls

downstand edge beam

5.000 outer radius

4.000 inner radius

1.650 rise

6.000

1 : 1 : 3/10 mm aggregate concrete to shell roof thickness at crown 75 mm

valley gutter screeded to falls

300 × 300 RC columns

300 300

valley beam

Typical expansion joint through columns

adjoining RC columns

joint filled externally with suitable mastic

copper expansion strip cast into columns

25 mm wide compressible material

Typical expansion joint through roof

copper clips at 450 % fixed to battens

roof finish

25 mm wide compressible material

copper capping secured with clips

75 mm wide × 100 mm high kerb

concrete shell roof

Fig V.12 Typical barrel vault details

158

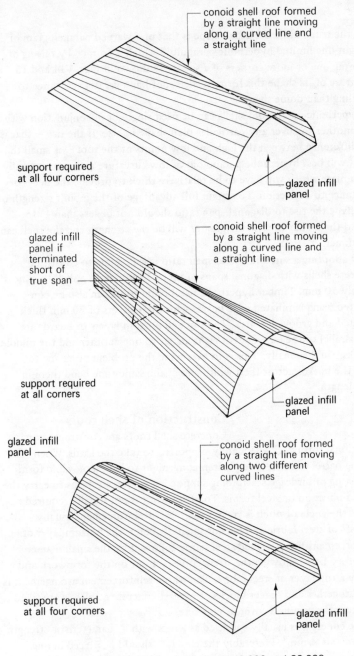

conoid shell roof formed by a straight line moving along a curved line and a straight line

support required at all four corners

glazed infill panel

glazed infill panel if terminated short of true span

conoid shell roof formed by a straight line moving along a curved line and a straight line

support required at all corners

glazed infill panel

glazed infill panel

conoid shell roof formed by a straight line moving along two different curved lines

support required at all four corners

glazed infill panel

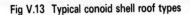

typical widths for all types between 12.000 and 30.000 spans very often made less than width

Fig V.13 Typical conoid shell roof types

than the true saddle the usual shape is that of a warped parallelogram or straight line limited hyperbolic paraboloid which is formed by raising or lowering one or more corners of a square as shown in Figs. V14 and 15. By virtue of its shape this form of shell roof has a greater resistance to buckling than dome shapes.

Hyperbolic paraboloid shells can be used singly or in conjunction with one another to cover any particular plan shape or size. If the rise — that is the difference between the high and low points of the roof — is small the result will be a hyperbolic paraboloid of low curvature acting structurally like a plate which will have to be relatively thick to provide the necessary resistance to deflection. To obtain full advantage of the inbuilt strength of the shape the rise to diagonal span ratio should not be less than 1 : 15; indeed the higher the rise the greater will be the strength and the shell can be thinner.

By adopting a suitable rise-to-span ratio it is possible to construct concrete shells with diagonal spans of up to 35.000 with a shell thickness of only 50 mm. Timber hyperbolic paraboloid roofs can also be constructed using laminated edge beams with three layers of 20 mm thick tongued and grooved boards. The top and bottom layers of boards are laid parallel to the edges but at right angles to one another and the middle layer is laid diagonally. This is to overcome the problem of having to twist the boards across their width and at the same time bend them in their length.

Construction of shell roofs

Concrete shell roofs are constructed on traditional formwork adequately supported to take the loads. When casting barrel vaults it is very often convenient to have a movable form consisting of birdcage scaffolding supporting curved steel ribs to carry the curved plywood or steel forms. Top formwork is not usually required unless the angle of pitch is greater than 45°. Reinforcement usually consists of steel fabric and bars of small diameter, the bottom layer of reinforcement being welded steel fabric followed by the small diameter trajectory bars following the stress curves set out on the formwork and finally a top layer of steel fabric. The whole reinforcement arrangement is wired together and spacer blocks of precast concrete or plastic are fixed to maintain the required cover of concrete.

The concrete is usually specified as a mix with a characteristic strength of 25 or 30 N/mm^2. Preferably the concrete should be placed in one operation in 1 m wide strips commencing at one end and running from edge beam to edge beam over the crown of the roof. A wet mix should be placed around the reinforcement followed by a floated drier mix. Thermal

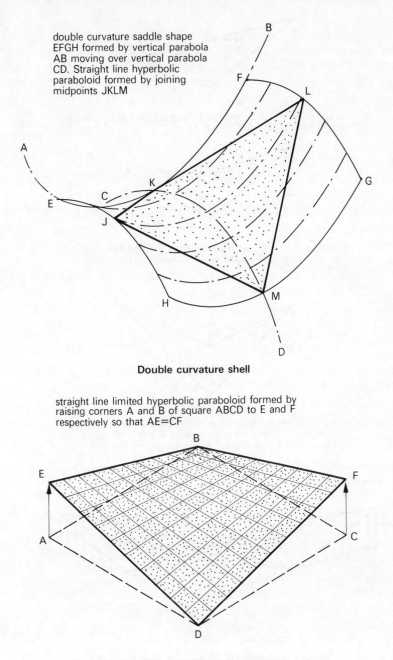

double curvature saddle shape
EFGH formed by vertical parabola
AB moving over vertical parabola
CD. Straight line hyperbolic
paraboloid formed by joining
midpoints JKLM

Double curvature shell

straight line limited hyperbolic paraboloid formed by
raising corners A and B of square ABCD to E and F
respectively so that AE=CF

Hyperbolic paraboloid shell

Fig V.14 Hyperbolic paraboloid roof principles

161

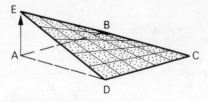

straight line limited hyperbolic paraboloid formed by raising
corner A of square ABCD to E

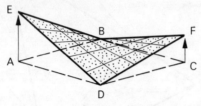

straight line limited hyperbolic paraboloid formed by raising
corners A and C of square ABCD to E and F respectively
so that AE≠CF

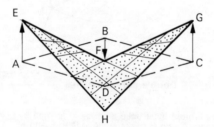

straight line limited hyperbolic paraboloid formed by raising
corners A and C and lowering corners B and D of square
ABCD to EFG and H respectively so that AE=CG and BF=DH

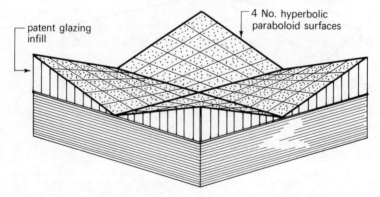

hyperbolic paraboloids combined to form single roof

Fig V.15 Typical hyperbolic paraboloid roof shapes

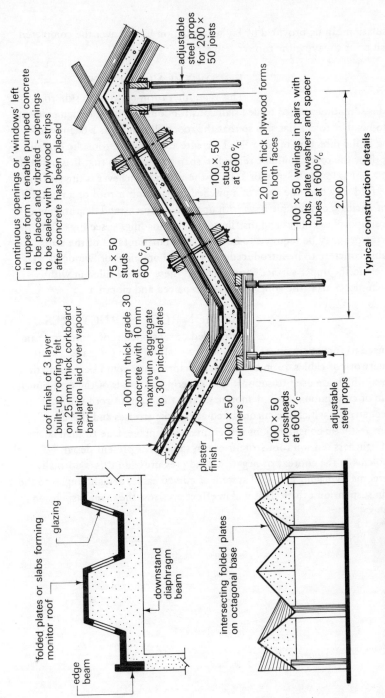

adjustable steel props for 200 × 50 joists

continuous openings or 'windows' left in upper form to enable pumped concrete to be placed and vibrated - openings to be sealed with plywood strips after concrete has been placed

100 × 50 studs at 600°c

20 mm thick plywood forms to both faces

100 × 50 walings in pairs with bolts, plate washers and spacer tubes at 600°c

75 × 50 studs at 600°c

2.000

roof finish of 3 layer built-up roofing felt on 25 mm thick corkboard insulation laid over vapour barrier

100 mm thick grade 30 concrete with 10 mm maximum aggregate to 30° pitched plates

100 × 50 runners

plaster finish

100 × 50 crossheads at 600°c

adjustable steel props

Typical construction details

folded plates or slabs forming monitor roof

glazing

edge beam

downstand diaphragm beam

intersecting folded plates on octagonal base

Fig V.16 Typical folded plate roof details

163

insulation can be provided by laying insulation blocks over the completed shell prior to laying the roof covering.

FOLDED PLATE ROOFS

This is another form of stressed skin roof and is sometimes called folded slab construction. The basic design concept is to bend or fold a flat slab so that the roof will behave as a beam spanning in the direction of the fold. To create an economic roof the overall depth of the roof should be related to span and width so that it is between 1/10 and 1/15 of the span or 1/10 of the width, whichever is the greater. The fold may take the form of a pitched roof, monitor roof or a multi-fold roof in single or multiple bays with upstand or downstand diaphragms at the supports to collect and distribute the slab loadings — see Fig. V.16. Formwork may be required to both top and bottom faces of the slabs. To enable concrete to be introduced and vibrated openings or 'windows' can be left in the upper surface formwork and these will be filled in with slip-in pieces after the concrete has been placed and vibrated.

TENSION ROOF STRUCTURES

Suspended or tensioned roof structures can be used to form permanent or temporary roofs and are generally a system or network of cables, or in the temporary form they could be pneumatic tubes, which are used to support roof covering materials of the traditional form or continuous sheet membranes. With this form of roof the only direct stresses which are encountered are tensile stresses and this apart from aesthetic considerations is their main advantage. Due to their shape and lightness tension roof structures can sometimes present design problems in the context of negative wind pressures and this is normally overcome by having a second system of curved cables at right angles to the main suspension cables. This will in effect prestress the main suspension cables.

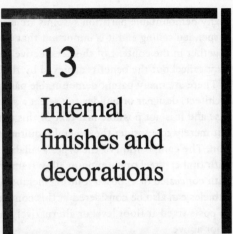

13
Internal finishes and decorations

Part VI
Finishes

Internal finishes which are normally associated with domestic dwellings such as brick walls, lightweight block walls, plastering, dry linings, floor finishes and coverings, fixed partitions and the various trims should have already been covered in the first two years of study. Advance level construction technology therefore usually concentrates on the finishes found in buildings such as offices, commercial and educational establishments. These finishes would include partitions, sliding doors and suspended ceilings.

PARTITIONS

A partition is similar to a wall in that it is a vertical construction dividing the internal space of a building but partitions are generally considered to be internal walls which are lightweight, non-load bearing, demountable or movable. They are used in buildings such as offices where it is desirable to have a system of internal division which can be altered to suit changes in usage without unacceptable costs, minimum of interference to services and ceilings. It should be noted that any non-load-bearing partition could if necessary be taken down and moved but in most cases moving partitions such as timber studwork with plasterboard linings will necessitate the replacement or repair of some of the materials and finishes involved. A true demountable partition can usually be taken down, moved and re-erected without any notable damage to the materials, components, finishes and surrounding parts of the building. Movable partitions usually take the form of a series of doors or door leaves hung to slide or slide and fold and these are described later under a separate heading.

165

Many demountable partition systems are used in conjunction with a suspended ceiling and it is important that both systems are considered together in the context of their respective main functions since one system may cancel out the benefits achieved by the other.

There are many patent demountable partition systems from which the architect, designer or builder can select a suitable system for any particular case and it is not possible in a text of this nature to analyse or list them all but merely to consider the general requirements of these types of partition. The composition of the types available range from glazed screens with timber, steel or aluminium alloy frames to the panel construction with concealed or exposed jointing members. Low level screens and toilet cubicles can also be considered in this context and these can be attached to posts fixed at floor level or alternatively suspended from the structural floor above.

When selecting or specifying a demountable partition the following points should be taken into account:

1. Fixing and stability — points to be considered are bottom fixing, top fixing and joints between sections. The two common methods of bottom fixing are the horizontal base unit which is very often hollow to receive services and the screw jack fixing, both methods are intended to be fixed above the finished floor level which should have a tolerance not exceeding 5 mm. The top fixing can be of a similar nature with a ceiling tolerance of the same magnitude but problems can arise when used in conjunction with a suspended ceiling since little or no pressure can be applied to the underside of the ceiling. A brace or panel could be fixed above the partition in the void over the ceiling to give the required stability or alternatively the partition could pass through the ceiling but this would reduce the flexibility of the two systems should a rearrangement be required at a later date. The joints between consecutive panels do not normally present any problems since they are an integral part of the design but unless they are adequately sealed the sound insulation and/or fire resistance properties of the partition could be seriously impaired.

2. Sound insulation — the degree of insulation required will depend largely upon the usage of rooms created by the partitions. The level of sound insulation which can be obtained will depend upon three factors: the density of the partition construction, degree of demountability and the continuity of the partition with the floor and ceiling. Generally, the less demountable a partition is the easier it will be to achieve an acceptable sound insulation level. Partitions

fixed between the structural floor and the structural ceiling usually result in good sound insulation whereas partitions erected between the floor and a suspended ceiling give poor sound insulation properties due mainly to the flanking sound path over the head of the partition. In the latter case two remedies are possible; firstly the partition could pass through the suspended ceiling, or alternatively a sound insulating panel could be inserted in the ceiling void directly over the partition below.

3. Fire resistance — the major restrictions to the choice of partition lie in the requirements of the Building Regulations Part B with regard to fire resistant properties, spread of flame, fire stopping and providing a suitable means of escape in case of fire. The weakest points in any system will be the openings and the seals at the foot and head. Provisions at openings are a matter of detail and adequate construction together with a suitable fire resistant door or fire resistant glazing. Fire stopping at the head of a partition used in conjunction with a suspended ceiling usually takes the form of a fire break panel fixed in the ceiling void directly above the partition below. Most proprietary demountable systems will give a half-hour or one hour fire resistance with a class 1 or class 0 spread of flame classification.

Typical demountable partition details are shown in Figs. VI.1 and 2.

SLIDING DOORS

Sliding doors may be used in all forms of buildings from the small garage to the large industrial structures. They may be incorporated into a design for any of the following reasons:

1. As an alternative to a swing door to conserve space or where it is not possible to install a swing door due to space restrictions.
2. Where heavy doors or doors in more than two leaves are to be used and where conventional hinges would be inadequate.
3. As a movable partition used in preference to a demountable partition because of the frequency with which it would be moved.

In most cases there is a choice of bottom or top sliding door gear which in the case of very heavy doors may have to be mechanically operated. Top gear can only be used if there is a suitable soffit, beam or lintel over the proposed opening which is strong enough to take the load of the gear, track and doors. In the case of bottom gear all the loads are transmitted to the floor. Top gear has the disadvantage of generally being noisier than its bottom gear counterpart but bottom gear either has an upstanding track which can be a hazard in terms of tripping or alternatively it can have a

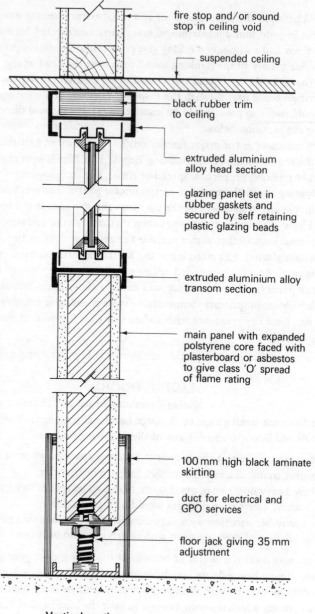

fire stop and/or sound
stop in ceiling void

suspended ceiling

black rubber trim
to ceiling

extruded aluminium
alloy head section

glazing panel set in
rubber gaskets and
secured by self retaining
plastic glazing beads

extruded aluminium alloy
transom section

main panel with expanded
polstyrene core faced with
plasterboard or asbestos
to give class 'O' spread
of flame rating

100 mm high black laminate
skirting

duct for electrical and
GPO services

floor jack giving 35 mm
adjustment

Vertical section

maximum panel height 3.657
panel thickness 54 mm overall
horizontal module 1.190

Fig VI.1 Typical demountable partition details 1 (Tenon Contracts Ltd)

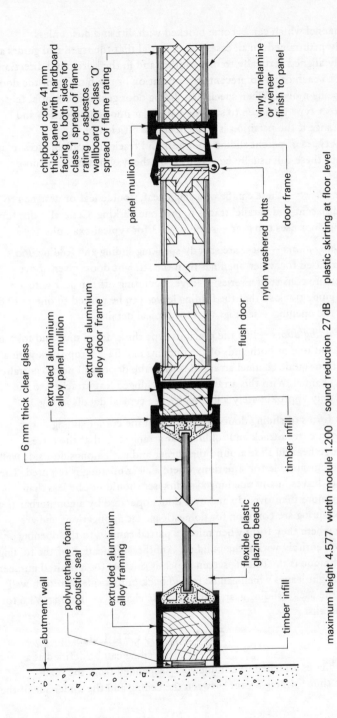

chipboard core 41 mm thick panel with hardboard facing to both sides for class 1 spread of flame rating or asbestos wallboard for class 'O' spread of flame rating

vinyl, melamine or veneer finish to panel

panel mullion

door frame

nylon washered butts

flush door

6 mm thick clear glass

extruded aluminium alloy panel mullion

extruded aluminium alloy door frame

timber infill

abutment wall

polyurethane foam acoustic seal

extruded aluminium alloy framing

flexible plastic glazing beads

timber infill

timber infill

maximum height 4.577   width module 1.200   sound reduction 27 dB   plastic skirting at floor level

services can be accommodated within chipboard core and along continuous base ducts.

Fig VI.2 Typical demountable partition details 2 (Venesta International Components Ltd)

sunk channel which can become blocked with dirt and dust unless regularly maintained. In all cases it is essential that the track and guides are perfectly aligned vertically to one another and in the horizontal direction parallel to each other to prevent stiff operation or the binding of the doors whilst being moved. When specifying sliding door gear it is essential that the correct type is chosen to suit the particular door combination and weight range if the partition is to be operated efficiently.

Many types of patent sliding door gear and track are available to suit all needs but these can usually be classified by the way in which the doors operate:

Straight sliding — these can be of a single leaf, double leaf or designed to run on adjacent and parallel tracks to give one parking. Generally this form of sliding door uses top gear — see Fig. VI.3 for typical example.

End-folding doors — these are strictly speaking sliding and folding doors which are used for wider openings than the straight doors given above. They usually consist of a series of leaves operating off top gear with a bottom guide track so that the folding leaves can be parked to one or both sides of the opening — see Fig. VI.4 for typical details.

Centre-folding doors — like the previous type these doors slide and fold to be parked at one or both ends of the opening and have a top track with a bottom floor guide channel arranged so that the doors will pivot centrally over the channel. With this arrangement the hinged leaf attached to the frame is only approximately a half-leaf — see typical details in Fig. VI.5.

Other forms of sliding doors include round the corner sliding doors which use a curved track and folding gear arranged so that the leaves hinged together will slide around the corner and park alongside a side wall, making them suitable for situations where a clear opening is required. The number of leaves in any one hinged sliding set should not be less than three nor more than five. Up and over doors operated by a counterbalance weight or spring are common for domestic garage applications and are available where they form when raised a partial canopy to the opening or are parked entirely within the building. Another application of the folding door technique is the folding screen which is suitable for a limited number of lightweight leaves which can be folded back onto a side or return wall without the use of sliding gear, track or guide channels — see Fig. VI.6 for typical details.

SUSPENDED CEILINGS

A suspended ceiling can be defined as a ceiling which is attached to a framework suspended from the main structure thus forming a void between the ceiling and the underside of the

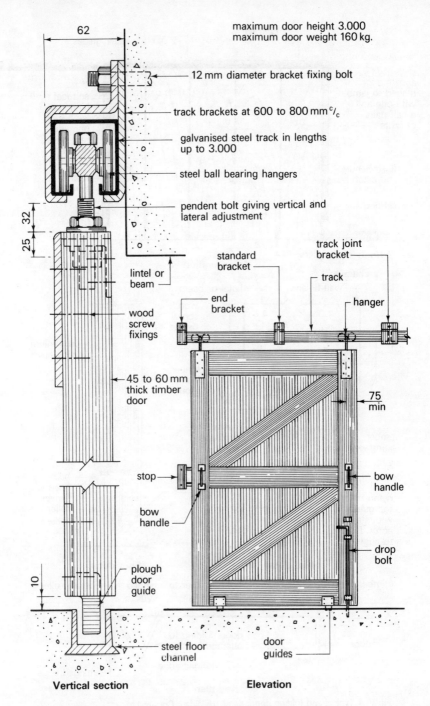

maximum door height 3.000
maximum door weight 160 kg.

— 12 mm diameter bracket fixing bolt

— track brackets at 600 to 800 mm $^c/_c$

— galvanised steel track in lengths up to 3.000

— steel ball bearing hangers

— pendent bolt giving vertical and lateral adjustment

62

32

25

lintel or beam

wood screw fixings

45 to 60 mm thick timber door

stop

bow handle

10

plough door guide

steel floor channel

Vertical section

track joint bracket

standard bracket

track

end bracket

hanger

75 min

bow handle

drop bolt

door guides

Elevation

Fig VI.3 Typical straight sliding door details (Hillaldam Coburn Ltd)

171

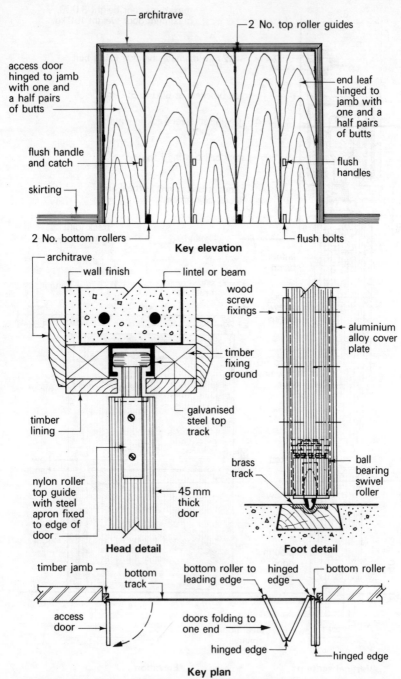

Fig VI.4 Typical end folding doors detail (Hillaldam Coburn Ltd)

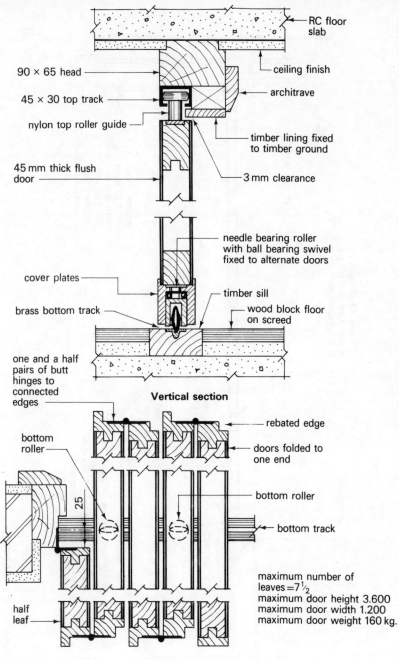

RC floor slab

ceiling finish

90 × 65 head

45 × 30 top track

nylon top roller guide

architrave

timber lining fixed to timber ground

45 mm thick flush door

3 mm clearance

needle bearing roller with ball bearing swivel fixed to alternate doors

cover plates

timber sill

brass bottom track

wood block floor on screed

one and a half pairs of butt hinges to connected edges

Vertical section

rebated edge

bottom roller

doors folded to one end

bottom roller

25

bottom track

maximum number of leaves = $7\frac{1}{2}$
maximum door height 3.600
maximum door width 1.200
maximum door weight 160 kg.

half leaf

Horizontal section

Fig VI.5 Typical centre folding sliding door details (Hillaldam Coburn Ltd)

173

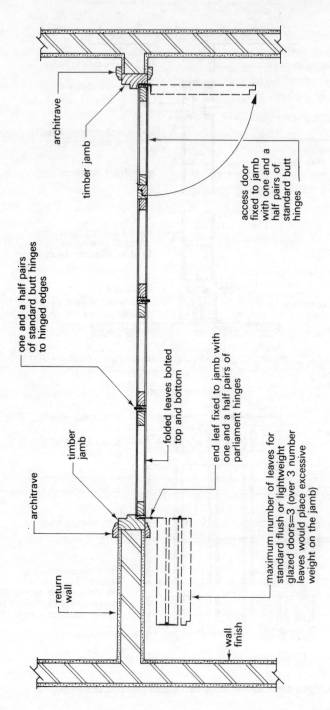

architrave

timber jamb

access door fixed to jamb with one and a half pairs of standard butt hinges

one and a half pairs of standard butt hinges to hinged edges

folded leaves bolted top and bottom

timber jamb

end leaf fixed to jamb with one and a half pairs of parliament hinges

architrave

return wall

maximum number of leaves for standard flush or lightweight glazed doors=3 (over 3 number leaves would place excessive weight on the jamb)

wall finish

Fig VI.6 Typical folding screen details

174

main structure. Ceilings which are fixed for example to lattice girders and trusses whilst forming a void are strictly speaking attached ceilings although they may be formed by using the same methods and materials as the suspended ceilings. The same argument can be applied to ceilings which are fixed to a framework of battens attached to the underside of the main structure.

The reasons for including a suspended ceiling system in a building design can be listed as follows:

1. Provide a finish to the underside of a structural floor or roof generally for purposes of concealment.
2. Create a void space suitable for housing and concealing services and light fittings.
3. Add to the sound and/or thermal insulation properties of the floor or roof above.
4. Provide a means of structural fire protection to steel beams supporting a concrete floor.
5. Provide a means of acoustic control in terms of absorption and reverberation.
6. Create a lower ceiling height to a particular room or space.

A suspended ceiling for whatever reason it has been specified and installed should fulfil the following requirements:

1. Easy to construct, repair, maintain and clean.
2. Conform with the requirements of the Building Regulations and in particular Regulations B2(a) and B2(b) which are concerned with the spread of flame over the surface and the limitations imposed on the use of certain materials.
3. Provide an adequate means of access for the maintenance of the suspension system and/or the maintenance of concealed services and light fittings.
4. Conform to a planning module which preferably should be based on the modular co-ordination recommendations set out in BS 4011 and BS 4330 which recommends a first preference module of 300 m.

There are many ways in which to classify suspended ceilings. For example, they can be classified by function such as an acoustic ceiling or by the materials involved, but one simple method of classification is to group the ceiling systems by their general method of construction thus:

1. Jointless ceilings.
2. Jointed ceilings.
3. Open ceilings.

Jointless ceilings: these are ceilings which, although suspended from the main structure, give the internal appearance of being a conventional ceiling. The final finish is usually of plaster applied in one or two coats to plasterboard or expanded metal lathing. Alternatively a jointless suspended ceiling could be formed by applying sprayed plaster or sprayed vermiculite-cement to an expanded metal background. Typical jointless ceiling details are shown in Fig. VI.7.

Jointed ceilings: these suspended ceilings are the most popular and common form because of their ease of assembly, ease of installation and ease of maintenance. They consist basically of a suspended metal framework to which the finish in a board or tile form are attached. The boards or tiles can be located on a series of tee bar supports with the supporting members exposed forming part of the general appearance or by using various spring clip devices or direct fixing the supporting members can be concealed. The common ceiling materials encountered are fibreboards, metal trays, fibre cement materials and plastic tiles or trays — see Fig. VI.8 for typical details.

Open ceilings: these suspended ceilings are largely decorative in function but by installing light fittings in the ceiling void they can act as a luminous ceiling. The format of these ceilings can be an open work grid with or without louvres or a series of closely spaced plates of polished steel or any other suitable material. The colour and texture of the sides and soffit of the ceiling void must be carefully designed if an effective system is to be achieved. It is possible to line these surfaces with an acoustic absorption material to provide another function to the arrangement — see typical details in Fig. VI.9.

DECORATIONS

The fundamentals of applying decorative finishes in the form of paint and wallpaper should have been covered in the elementary studies of a typical four-year course in construction technology and therefore study at advance level concentrates on deeper aspects of the basic principles particularly in the context of painting timber and metals.

Painting timber

Timber can be painted to prevent decay in the material by forming a barrier to the penetration of moisture thus giving rise to the conditions necessary for fungal attack to begin, or alternatively timber may be painted mainly to impart colour. Timber

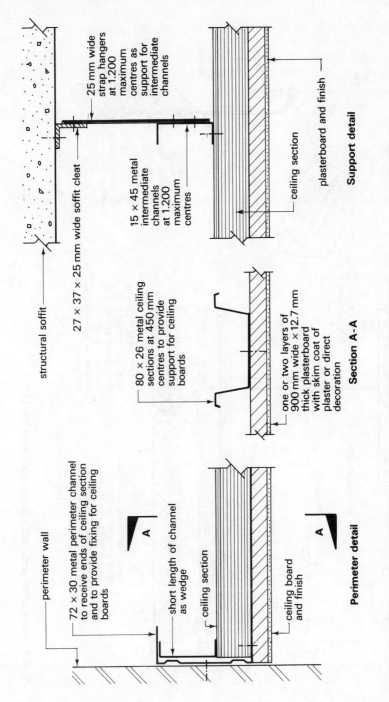

structural soffit

27 × 37 × 25 mm wide soffit cleat

25 mm wide strap hangers at 1.200 maximum centres as support for intermediate channels

15 × 45 metal intermediate channels at 1.200 maximum centres

plasterboard and finish

ceiling section

Support detail

80 × 26 metal ceiling sections at 450 mm centres to provide support for ceiling boards

one or two layers of 900 mm wide × 12.7 mm thick plasterboard with skim coat of plaster or direct decoration

Section A-A

perimeter wall

72 × 30 metal perimeter channel to receive ends of ceiling section and to provide fixing for ceiling boards

A

short length of channel as wedge

ceiling section

ceiling board and finish

A

Perimeter detail

Fig VI.7 Typical jointless suspended ceiling details (Gyproc M/F suspended ceiling system)

177

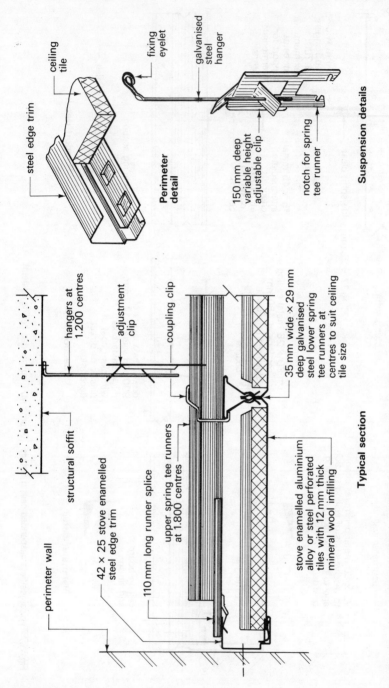

Perimeter detail

ceiling tile

steel edge trim

fixing eyelet

galvanised steel hanger

150 mm deep variable height adjustable clip

notch for spring tee runner

Suspension details

perimeter wall

structural soffit

hangers at 1.200 centres

adjustment clip

coupling clip

42 × 25 stove enamelled steel edge trim

110 mm long runner splice

upper spring tee runners at 1.800 centres

35 mm wide × 29 mm deep galvanised steel lower spring tee runners at centres to suit ceiling tile size

stove enamelled aluminium alloy or steel perforated tiles with 12 mm thick mineral wool infilling

Typical section

Fig VI.8 Typical jointed suspended ceiling details (Dampa (UK) Ltd)

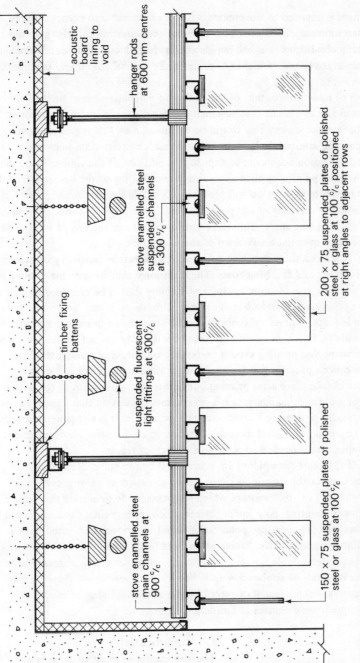

acoustic board lining to void

hanger rods at 600 mm centres

stove enamelled steel suspended channels at 300 %c

200 × 75 suspended plates of polished steel or glass at 100 %c positioned at right angles to adjacent rows

timber fixing battens

suspended fluorescent light fittings at 300 %c

stove enamelled steel main channels at 900 %c

150 × 75 suspended plates of polished steel or glass at 100 %c

Fig VI.9 Typical open suspended ceiling details

179

which is exposed to the elements is more vulnerable to eventual decay than internal joinery items but internal condensation can also give rise to damp conditions. Timber can also be protected with clear water-repellent preservatives and varnishes to preserve the natural colour and texture of the timber.

The moisture content of the timber to which paint is to be applied should be as near to that at which it will stabilise in its final condition. For internal joinery this would be within an 8 to 12% region depending upon the internal design temperature and for external timber within a 15 to 18% region according to exposure conditions. If the timber is too dry when the paint is applied any subsequent swelling of the base material due to moisture absorption will place unacceptable stresses on the paint film leading to cracking of the paint barrier. Conversely wet timber drying out after the paint application can result in blistering, opening of joints and the consequential breakdown of the paint film.

New work should receive a four-coat application consisting of primer, undercoat and finishing coats. The traditional shellac knotting applied to reduce the risk of resin leaking and staining should be carried out prior to the primer application but in external conditions it may prove to be inadequate. Therefore timber selected for exposed or external conditions should be of a high quality without or with only a small amount of knots. Knotting and priming should preferably be carried out under the ideal conditions prevailing at the place of manufacture. Filling and stopping should take place after priming but before the application of the under-coat and finishing coats. It is a general misconception that a good key is necessary to achieve a satisfactory paint surface. It is not penetration that is required but a good molecular attraction of the paint binder to the timber that is needed to obtain a satisfactory result.

If the protective paint film is breached by moisture, decay can take place under the paint coverings; indeed the coats of paint prevent the drying out of any moisture which has managed to penetrate the timber prior to painting. Any joinery items which in their final situation could be susceptible to moisture penetration should therefore be treated with a paintable preservative. Preservation treatments include diffusion treatment using water-soluble borates, water-borne preservatives and organic solvents. The methods of application and chemical composition of these timber preservatives are usually covered by the science syllabus of most comprehensive courses in building.

Painting metals

The application of paint to provide a protective coat and for decorative purposes is normally confined to iron and steel since most non-ferrous metals are left to oxidise and form their own natural protective coating. Corrosion of ferrous metals is a natural process which can cause disfiguration and possible failure of a metal component. Two aspects of painting these metals must be considered:

1. Initial preparation and paint application.
2. Maintenance of protective paint.

Preparation: the basic requirement is to prepare the metal surface adequately to receive the primer and subsequent coats of paint since the mill scale, rust, oil, grease and dirt which are frequently found on metals are not suitable as a base for the applications of paint. Suitable preparation treatments are:

1. Shot and grit blasting — effective method which can be carried out on site although it is usually considered to be a factory process.
2. Phosphating and pickling — factory process involving immersion of the metal component in hot acid solutions to remove rust and scale.
3. Degreasing — factory process involving washing or immersion using organic solvents, emulsions or hot alkali solutions followed by washing. Very often used as a preliminary treatment to phosphating.
4. Mechanical — site or factory process using hand held or powered tools such as hammers, chisels, brushes and scrapers. To be effective these forms of surface preparation must be thorough.
5. Flame cleaning — an oxy-acetylene flame used in conjunction with hand-held tools for removing existing coats of paint or loose scale and rust.

It is essential that in conjunction with the above preparation processes the correct primer is specified and used to obtain a satisfactory result. As with the painting of timber the types of primers and paints available together with the methods of application are usually contained in the science syllabus.

Maintenance: to protect steel in the long term a rigid schedule of painting maintenance must be worked out and invoked. This is particularly true of the many areas where access is difficult such as pipes fixed close to a wall and adjacent members in a lattice truss. Access for future maintenance is an aspect of building which should be considered during the design and construction stages not only in the context of metal components but to any part of a building which will require future maintenance and/or inspection. To take full advantage of the structural properties of steelwork and at the same time avoid the maintenance problems, weathering steels could be considered for exposed conditions.

Weathering steels

These do not rust or corrode as normal steels but interact with the atmosphere to produce a layer of sealing oxides. The colour of this protective layer will vary from a lightish brown to a dark purple grey depending on the degree of exposure, amount and type of pollution in the atmosphere and orientation. The best known weathering steel is called 'Cor-Ten' which is derived from the fact that it is CORrosion resistant with a high TENsile strength. Cor-Ten is not a stainless steel but a low alloy steel with a lower proportion of non-ferrous metals than stainless steel which makes it a dearer material than mild steel but cheaper than traditional stainless steels.

Weathering steels can be used as a substitute for other steels except in wet situations, marine works or in areas of high pollution unless protected with paint or similar protective applications which defeats the main objective of using this material. Jointing can be by welding or friction grip bolts. It must be noted that the protective coating will not form on weathering steels in internal situations since the formation of this coat is a natural process of the wet and dry cycles encountered with ordinary weather conditions.

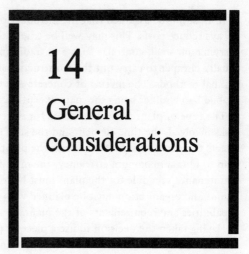

14
General considerations

Part VII
Builders' plant

The general aim of building is to produce a structure of reasonable cost and sound workmanship within an acceptable time period. To achieve this time period and in many cases to overcome a shortage of suitable manpower the mechanisation of many building activities must be considered. The items of plant now available to building contractors is very extensive ranging from simple hand tools to very expensive equipment undertaking tasks beyond the capabilities of manual labour. In a text of this nature it is only possible to consider the general classes of plant and their uses; for a full analysis of the many variations with the different classifications, students are advised to consult the many text-books and catalogues devoted entirely to contractors' plant.

The main reasons for electing to use items of plant can be enumerated as follows:

1. Increase rate of output.
2. Reduce overall building costs.
3. Carry out activities which cannot be done manually or to do them more economically.
4. Eliminate heavy manual work thus reducing fatigue and increasing the productivity of manual workers.
5. Maintain a planned rate of production where there is a shortage of either skilled or unskilled labour.
6. Maintain the high standards often required by present-day designs and specifications especially when concerned with structural engineering works.

It must not be assumed that the introduction of plant to a contract will always reduce costs. This may well be true with large contracts but when carrying out small contracts such as a traditionally built one off house it is usually cheaper to carry out the constructional operations by traditional manual methods. The mixing of concrete and cement mortar using a small mobile batch mixer being the main exception.

The type of plant to be considered for selection will depend upon the tasks involved, the time element and the staff available. The person who selects the plant must be competent, the plant operator must be a trained man to obtain maximum efficiency, the manufacturer's recommended maintenance schedule for the plant must be followed and above all the site layout and organisation must be planned with a knowledge of the capabilities and requirements of the plant.

Having taken the decision to use a piece or pieces of plant the contractor now has the choice of buying, hiring or a combination of the two. The advantages of buying plant can be listed as follows:

1. Plant is available when required.
2. Cost of idle time caused by inclement weather, work being behind planned programme or delay in deliveries of materials will generally be less on owned plant than hired plant.
3. Builder can apportion the plant costs to the various contracts using the plant by his own chosen method.

The advantages of hiring plant can be enumerated as follows:

1. Plant can be hired as required and for short periods.
2. Hire firms are responsible for repairs and replacements.
3. Contractor is not left with expensive plant items on his hands after the completion of the contract.
4. Hire rates can include operator, fuel and oil.

Quotations and conditions of hiring plant can be obtained from a plant hiring company, a full list of companies belonging to the Contractors' Plant Association is contained in the CPA Annual Handbook. This Association was formed in 1941 to represent the plant hire industry in the United Kingdom to negotiate general terms and conditions of hiring plant, to give advice and to promote a high standard of efficiency in the services given by its members. Against plant hire rates and conditions a contractor will have to compare the cost of buying and owning a similar piece of plant. A comparison of costs can be calculated by the simple straight line method or by treating the plant as an investment and charging interest on the capital outlay.

Straight line method:

Capital cost of plant	£4 500.00
Expected useful life	5 years
Yearly working, say 75% of total	

year's working hours $= 50 \text{ weeks} \times 40 \text{ hours} \times \dfrac{75}{100}$

$= 1\ 500$ hours per year

Assuming a resale value at the end of the 5-year period of £750.00

Annual depreciation $= \dfrac{4\ 500 - 750}{5} = £750.00$

Therefore hourly depreciation $= \dfrac{750}{1\ 500} = 50\text{p}$

Net cost per hour	= 50p
Add 2% for insurance, etc.	= 1p
Add 10% for maintenance	= 5p
Total	56p per hour

To the above figure must be added the running costs which would include fuel, operator's wages and overheads.

Interest on capital outlay method:

Capital cost of plant	£4 500.00
Interest on capital – 6% for 5 years	1 350.00
	£5 850.00
Deduct resale value	750.00
	£5 100.00
Add 2% of capital cost for insurance, etc.	90.00
Add 10% of capital cost for maintenance	450.00
	£5 640.00

Therefore cost per hour $= \dfrac{5\ 640}{5 \times 1\ 500} = 75\text{p}$

To the above hourly rate must be added the running costs as given for the straight line method. This second method gives a more accurate figure to the actual cost of owning an item of plant than the straight line method and for this reason it is widely used.

Vehicles such as lorries and vans are usually costed on a straight annual depreciation of the yearly book value thus:

Cost of vehicle	£2 500.00
Estimated useful life, 5 years	
Annual depreciation, 30%	
Capital cost	£2 500.00
30% depreciation	750.00

Value after 1st year	1 750.00
30% depreciation	525.00
Value after 2nd year	1 225.00
30% depreciation	367.50
Value after 3rd year	857.50
30% depreciation	257.25
Value after 4th year	600.25
30% depreciation	180.08
Value at 5th year	£420.17

The percentages of insurance and maintenance together with the running costs can now be added on a yearly basis taking the new book values for each year, or alternatively the above additional costs could be averaged over the five-year period.

To be an economic proposition large items of plant need to be employed continuously and not left idle for considerable periods of time. Careful maintenance of all forms of plant is of the utmost importance not only to increase the working life of a piece of plant but if a plant failure occurs on site it can cause serious delays and disruptions of the programme and this in turn can affect the company's future planning. To reduce the risk of plant breakdown a trained and skilled operator should be employed to be responsible for the running, cleaning and daily maintenance of any form of machinery. Time for the machine operator to carry out these tasks must be allowed for in the site programme and the daily work schedules.

On a large contract where a number of machines are to be employed a full-time skilled mechanic could be engaged to be responsible for running repairs and recommended preventive maintenance. Such tasks would include:

1. Checking oil levels – daily.
2. Greasing – daily or after each shift.
3. Checking engine sump levels – after 100 hours running time.
4. Checking gearbox levels – after 1 200 to 1 500 hours running time.
5. Checking tyre pressures – daily.
6. Inspecting chains and ropes – daily.

As soon as a particular item of plant has finished its work on site it should be returned to the company's main plant yard so that it can be re-allocated to another contract. On its return to the main plant yard an item of plant should be inspected and tested so that any necessary repairs, replacements and maintenance can be carried out before it is re-employed

186

on another site. A record of the machine's history should be accurately kept and should accompany the machine wherever it is employed so that the record can be kept up-to-date.

The soil conditions and modes of access to a site will often influence the choice of plant items which could be considered for a particular task. Congested town sites may severely limit the use of many types of machinery and/or plant. If the proposed structure occupies the whole of the site it could eliminate the use of large batch concrete mixers, dumpers and cement storage silos. Wet sites usually require plant equipped with caterpillar tracks whereas dry sites are suitable for tracked and wheeled vehicles or power units. On housing sites it is common practice to construct the estate roads at an early stage in the contract to provide the firm access routes for mobile plant and hardstanding for static plant such as cement mixers. Sloping sites are usually unsuitable for rail-mounted cranes but these cranes operating on the perimeter of a building are more versatile than the static cranes. The heights and proximity of adjacent structures or buildings may limit the use of a horizontal jib crane and may even dictate the use of a crane with a luffing jib.

For accurate pricing of a Bill of Quantities careful consideration of all plant requirements must be undertaken at the pre-tender stage taking into account plant types, plant numbers and personnel needed. If the tender is successful a detailed programme should be prepared in liaison with all those to be concerned in the supervision of the contract so that the correct sequence of operations is planned and an economic balance of labour and machines is obtained.

Apart from the factors discussed above consideration must also be given to safety and noise emission requirements when selecting items of plant for use on a particular contract. The aspects of safety which must be legally provided are contained in various Acts of Parliament and Statutory Instruments such as the Health and Safety at Work, etc., Act 1974 and the Construction Regulations. These requirements will be considered in the following chapters devoted to various classifications of plant.

Although reaction to noise is basically subjective excessive noise can damage a person's health and/or hearing, it can also cause disturbance to working and living environments. Under the Health and Safety at Work, etc., Act 1974 provision is made for the protection of workers against noise. The maximum safe daily dosages of noise for the unprotected human ear are given in the recommendations of the relevant Department of Employment Code of Practice for reducing the exposure of employed persons to noise. If these daily dosages are likely to be exceeded the remedies available are the issue of suitable ear protectors to the workers or to use quieter plant or processes.

187

Local authorities have powers under the Control of Pollution Act 1974 to protect the community against noise. Part III of this Act gives the local authority the power to specify its own requirements as to the limit of construction noise acceptable by serving a notice which may specify:

1. Plant or machinery which is, or is not, to be used.
2. Hours during which the works may be carried out.
3. Level of noise emitted from the premises in question or from any specific point on these premises or the level of noise which may be emitted during specified hours.

The local authority is also allowed to make provisions in these notices for any change of circumstance.

A contractor will therefore need to know the local authority requirements as to noise restrictions at the pre-tender stage to enable him to select the right plant and/or process to be employed. Methods for predicting site noise are given in BS 5228 Code of Practice, Noise Control on Construction and Demolition sites. This Code outlines the working of the controlling Act, gives noise outputs of some 150 plant items and also gives the procedure for predicting site noise in terms of equivalent continuous sound levels as well as maximum sound levels. These are the levels which local authorities use when specifying construction noise limits.

15

Small powered plant

A precise definition of builders' small powered plant is not possible since the term 'small' is relative. For example, it is possible to have small cranes but these when compared with a hand-held electric drill would be considered as large plant items. Generally therefore small plant can be considered to be hand held or operated power tools with their attendant power sources such as a compressor for pneumatic tools, one of the main exceptions being the small static pumps used in conjunction with shallow excavations.

Most hand-held power tools are operated by electricity or compressed air either to rotate the tool or drive it by percussion. Some of these tools are also designed to act as rotary/percussion tools. Generally the pneumatic tools are used for the heavier work and have the advantage that they will not burn out if a rotary tool stalls under load. Electrically driven tools are however relatively quiet since there is no exhaust noise and can be used in confined spaces because there are no exhaust fumes.

ELECTRIC HAND TOOLS

The most common hand-held tool is the electric drill for boring holes into timber, masonry and metals using twist drills. A wide range of twist drill capacities are available with single- or two-speed motors for general purpose work, the low speed being used for boring into or through timber. Chuck capacities generally available range from 6 to 30 mm for twist drills suitable for metals. Dual purpose electric drills are very versatile in that the rotary motion can be combined with or converted into a powerful but rapid percussion motion making the tool

189

suitable for boring into concrete providing that special tungsten carbide-tipped drills are used — see Fig. VII.1.

A variation of the basic electric motor power units is the electric hammer which is used for cutting and chasing work where the hammer delivers powerful blows at a slower rate than the percussion drill described above. Another variation is the electric screwdriver which has an adjustable and sensitive clutch which will only operate when the screwdriver bit is in contact with the screw head and will slip when a predetermined tension has been reached when the screw has been driven in. Various portable electric woodworking tools such as circular saws, jigsaws, sanders, planes and routers are also available and suitable for site use.

Electric hand-held tools should preferably operate off a reduced voltage supply of 110 V and should conform to the recommendations of BS 2769. All these tools should be earthed unless they bear the 'two squares' symbol indicating that they are 'All Insulated' or 'Double Insulated' and therefore have their own built-in safety system. Plugs and couplers should comply with BS 4343 so that they are not interchangeable and cannot be connected to the wrong voltage supply. Protective guards and any recommended protective clothing such as goggles and ear protection should be used as instructed by the manufacturers and as laid down in the Construction (General Provisions) Regulations 1961 and in accordance with the Protection of Eyes Regulations 1974. Electric power tools must never be switched off whilst under load since this could cause the motor to become overstrained and burnt out. If electrical equipment is being used on a site the Electricity Regulations and relevant first-aid placards should be displayed in a prominent position.

PNEUMATIC TOOLS

These tools need a supply of compressed air as their power source and on building sites this is generally in the form of a mobile compressor powered by a diesel, petrol or electric motor, the most common power unit being the diesel engine. Compressors for building works are usually of the piston or reciprocating type where the air is drawn into a cylinder and compressed by a single stroke of the piston; alternatively two-stage compressors are available where the air is compressed to an intermediate level in a large cylinder before passing to a smaller cylinder where the final pressure is obtained. Air receivers are usually incorporated with and mounted on the chassis to provide a constant source of pressure for the air lines and to minimise losses due to pressure fluctuations of the compressor and/or frictional losses due to the air pulsating through the distribution hoses.

One of the most common pneumatic tools used in building is the

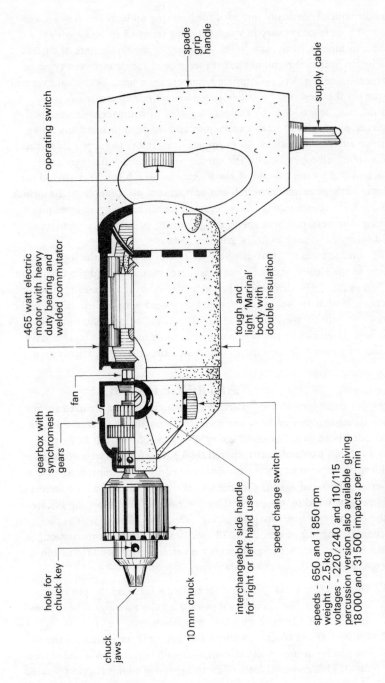

spade grip handle

supply cable

operating switch

465 watt electric motor with heavy duty bearing and welded commutator

tough and light 'Marinal' body with double insulation

gearbox with synchromesh gears

fan

hole for chuck key

chuck jaws

10 mm chuck

interchangeable side handle for right or left hand use

speed change switch

speeds – 650 and 1850 rpm
weight – 2.5 kg
voltages – 220/240 and 110/115
percussion version also available giving
18 000 and 31 500 impacts per min

Fig VII.1 Typical electric drill (Stanley-Bridges Ltd)

191

breaker which is basically intended for breaking up hard surfaces such as roads. These breakers vary in weight ranging from 15 to 40 kg with air consumptions of 10 to 20 m^3/min. A variety of breaker points or cutters can be fitted into the end of the breaker tool to tackle different types of surfaces — see Fig. VII.2. Chipping hammers are a small lightweight version of the breaker described above having a low air consumption of 0.5 m^3/min or less. Backfill tampers are used to compact the loose spoil returned as backfill in small excavations and weigh approximately 23 kg with an air consumption of 10 m^3/min. Compressors to supply air to these tools are usually specified by the number of air hoses which can be attached and by the volume of compressed air which can be delivered per minute. Other equipment which can be operated by compressed air include vibrators for consolidating concrete, small trench sheeting or sheet pile driving hammers, concrete spraying equipment, paint sprayers and hand-held rotary tools such as drills, grinders and saws.

Pneumatic tools are generally very noisy and in view of the legal requirements of the Health and Safety at Work, etc., Act 1974 and the Control of Pollution Act 1974 these tools should be fitted with a suitable muffler or silencer. Models are being designed and produced with built-in silencers fitted not only on the tool holder but also on the compressor unit. The mufflers which can be fitted to pneumatic tools such as the breaker have no loss of power effect on the tool provided the correct muffler is used.

CARTRIDGE HAMMERS

Cartridge hammers or guns are used for the quick fixing together of components or for firing into a surface a pin with a threaded head to act as a bolt fixing. The gun actuates a 0.22 cartridge which drives a hardened austempered steel pin into an unprepared surface. Holding power is basically mechanical, caused by the compression of the material penetrated against the shank of the pin, although with concrete the heat generated by the penetrating force of the pin causes the silicates in the concrete to fuse into glass giving a chemical bond as well as the mechanical holding power. BS 4078 specifies the design, construction, safety and performance requirements for cartridge-operated fixing tools, this standard also defines two basic types of tool:

1. Direct-acting tools — in which the driving force on the pin comes directly from the compressed gases from the cartridge. These tools are high velocity guns with high muzzle energy.
2. Indirect-acting tools — in which the driving force is transmitted to the pin by means of an intervening piston with limited axial movement. This common form of cartridge-fixing tool is trigger operated

Breaker   **Moil point**   **Chisel point**

Tarmac cutter   **Clay spade**   **Chipping hammer and point**

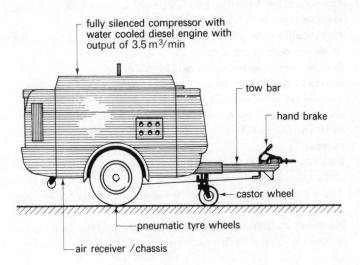

fully silenced compressor with
water cooled diesel engine with
output of 3.5 m³/min

tow bar

hand brake

castor wheel

pneumatic tyre wheels

air receiver /chassis

Typical 2 tool silenced compressor

Fig VII.2 Typical pneumatic tools and compressor

having a relatively low velocity and muzzle energy, a typical example being shown in Fig. VII.3.

Hammer-actuated fixing tools working on the piston principle are also available, the tool being hand held and struck with a hammer to fire the cartridge.

The cartridges and pins are designed for use with a particular model of gun and should under no circumstances be interchanged between models or different makes of gun. Cartridges are produced in seven strengths ranging from extra low to extra high and for identification purposes are colour coded in accordance with BS 4078. A low strength cartridge should be first used for a test firing, gradually increasing the strength until a satisfactory result is obtained. If an over-strength cartridge is used it could cause the pin to pass right through the base material. Fixing pins near to edges can be dangerous due to the pin deflecting towards the free edge by following the path of low resistance. Minimum edge distances recommended for concrete are 100 mm using a direct-acting tool and 50 mm using an indirect-acting tool; for fixing into steel the minimum recommended distance is 15 mm using any type of fixing tool.

The safety and maintenance aspects of these fixing tools cannot be overstressed and in particular consideration must be given to the following points:

1. Pins should not be driven into brittle or hard materials such as vitreous-faced bricks, cast iron and marble.
2. Pins should not be driven into base materials where there is a danger of the pin passing through.
3. Pins should not be fired into existing holes.
4. During firing the tool should be held at right angles to the surface with the whole of the splinter guard flush with the surface of the base material. Some tools are designed so that they will not fire if the angle of the axis of the gun with the perpendicular exceeds 7°.
5. Operatives under the age of 18 years should not be allowed to use cartridge-operated tools.
6. All operators should have a test for colour blindness before being allowed to use these fixing tools.
7. No one apart from the operator and his assistant should be in the immediate vicinity of firing to avoid accidents due to richochet, splintering or re-emergence of pins.
8. Operators should receive instructions as to the manufacturers' recommended method of loading, firing and the action to be taken in cases of misfiring.

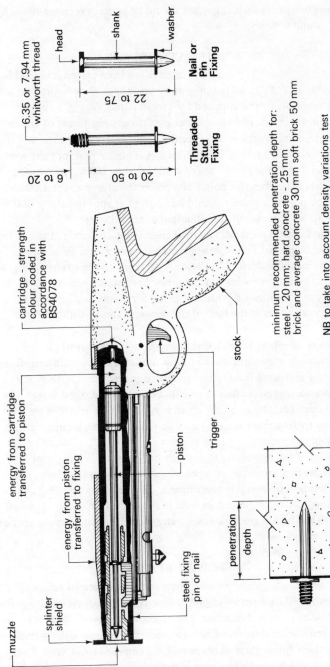

head
shank
washer

Nail or Pin Fixing

22 to 75

6.35 or 7.94 mm whitworth thread

6 to 20

20 to 50

Threaded Stud Fixing

cartridge - strength colour coded in accordance with BS4078

energy from cartridge transferred to piston

energy from piston transferred to fixing

muzzle

splinter shield

steel fixing pin or nail

piston

trigger

stock

minimum recommended penetration depth for: steel - 20 mm; hard concrete - 25 mm brick and average concrete 30 mm soft brick 50 mm

NB to take into account density variations test firings should always be carried out.

penetration depth

Fig VII.3 Typical cartridge hammer (Douglas Kane Group Ltd)

195

9. Protective items such as goggles should be worn as recommended by the manufacturers.

VIBRATORS

After concrete has been placed it should be consolidated either by hand tamping or by using special vibrators. The power for vibrators can be supplied by a small petrol engine, an electric motor and in some cases by compressed air. Three basic forms of vibrator are used in building works — these are the poker vibrator, vibration tampers and clamp vibrators. Poker vibrators are immersed into the wet concrete and due to their high rate of vibration they induce the concrete to consolidate. The effective radius of a poker vibrator is about 1.000, therefore the poker or pokers should be inserted at approximately 600 mm centres to achieve an overall consolidation of the concrete.

Vibration tampers are small vibrating engines which are fixed to the top of a tamping board for consolidating concrete pavings and slabs. Clamp vibrators are a similar device but these are attached to the external sides of formwork to vibrate the whole of the form. Care must be taken when using this type of vibrator to ensure that the formwork has sufficient in-built strength to resist the load of the concrete and to withstand the vibrations.

In concrete members which are thin and heavily reinforced careful vibration will cause the concrete to follow uniformly around the reinforcement and this increased fluidity due to vibration will occur with mixes which under normal circumstances would be considered too dry for reinforced concrete. Owing to the greater consolidation achieved by vibration up to 10% more material may be required when compared with hand-tamped concrete.

Separation of the aggregates can be caused by over vibrating a mix; therefore vibration should be stopped when the excess water rises to the surface. Vibration of concrete saves time and labour in the placing and consolidating of concrete but does not always result in a saving in overall costs due to the high formwork costs, extra materials costs and the cost of providing the necessary plant.

POWER FLOATS

Power floats are hand-operated rotary machines powered by a petrol engine or an electric motor which drives the revolving blades or a revolving disc. The objective of power floats is to produce a smooth level surface finish to concrete beds and slabs suitable to receive the floor finish without the need for a cement/sand screed. The surface finish which can be obtained is comparable with that achieved by craftsmen using hand trowels but takes only one-sixth of the time giving a considerable saving in both time and money. Most power floats can be fitted with either a revolving disc or blade head and are generally interchangeable.

The surfacing disc is used for surface planing after the concrete has been vibrated and will erase any transverse tamping line marks left by a vibrator beam as well as filling in any small cavities in the concrete surface. The revolving blades are used after the disc planing operation to provide the finishing and polishing which can usually be achieved with two passings. The time at which disc planing can be started is difficult to specify, depending on factors such as the workability of the concrete, temperature, relative humidity and the weight of the machine to be used. Experience is usually the best judge but as a guide if imprints of not more than 2 to 4 mm deep are made in the concrete when walked upon it is generally suitable for disc planing. It should be noted that if a suitable surface can be produced by traditional concrete placing method the disc planing operation prior to rotary blade finishing is often omitted. Generally blade finishing can be commenced once the surface water has evaporated, a typical power float being shown in Fig. VII.4. Power floats can also be used for finishing concrete floors with a granolithic or similar topping.

PUMPS

Pumps are one of the most important items of small plant for the building contractor since they must be reliable in all conditions, easy to maintain, easily transported and efficient. The basic function of a pump is to move liquids vertically or horizontally or in a combination of the two directions. Before selecting a pump a builder must consider what task the pump is to perform and this could be any of the following:

1. Keeping excavations free from water.
2. Lowering the water table to a reasonable depth.
3. Moving large quantities of water such as the dewatering of a cofferdam.
4. Supplying water for general purposes.

Having defined the task the pump is required to carry out, the next step is to choose a suitable pump taking into account the following factors:

1. Volume of water to be moved.
2. Rate at which water is to be pumped.
3. Height of pumping which is the vertical distance from the level of the water to the pump and is usually referred to as 'suction lift' or

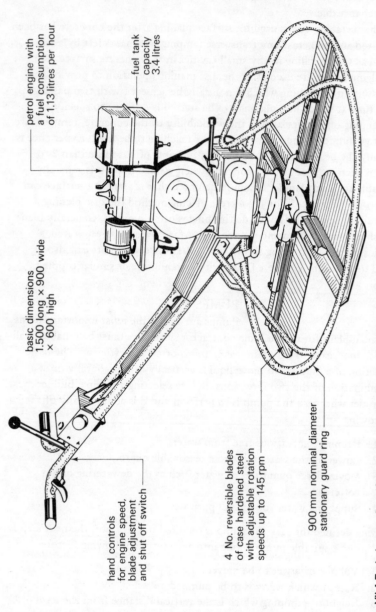

petrol engine with
a fuel consumption
of 1.13 litres per hour

fuel tank
capacity
3.4 litres

basic dimensions
1.500 long × 900 wide
× 600 high

hand controls
for engine speed,
blade adjustment
and shut off switch

4 No. reversible blades
of case hardened steel
with adjustable rotation
speeds up to 145 rpm

900 mm nominal diameter
stationary guard ring

Fig VII.4 Typical power float (Construction Equipment and Machinery (Gt. Britain) Ltd)

head and should be the shortest distance practicable to obtain economic pumping.

4. Height and distance to outfall or discharge point and is usually called the 'delivery head'.

5. Loss due to friction in the length of hose or pipe which increases as the diameter decreases. In many pumps the suction and delivery hoses are marginally larger in diameter than the pump inlet to reduce these frictional losses.

6. Power source for pump which can be a petrol engine, diesel engine or an electric motor. Pumps powered by compressed air are available but these are unusual on general building contracts.

Pumps in common use for general building works can be classified as follows:

1. Centrifugal.
2. Displacement.
3. Submersible.

Centrifugal pumps: classed as normal or self-priming and consist of a rotary impeller which revolves at high speed forcing the water to the sides of the impeller chamber thus creating a vortex which sucks air out of the suction hose. Atmospheric pressure acting on the surface of the water to be pumped causes the water to rise into the pump initiating the pumping operation. Normal centrifugal pumps are easy to maintain but they require priming with water at the commencement of each pumping operation. Where continuous pumping is required such as in a basement excavation a self-priming pump should be specified. These pumps have a reserve supply of water in the impeller chamber so that if the pump runs dry the reserve water supply will remain in the chamber to reactivate the pumping sequence if the water level rises in the area being pumped.

Displacement pumps: either reciprocating or diaphragm pumps. Reciprocating pumps work by the action of a piston or ram moving within a cylinder. The action of the piston draws water into the cylinder with one stroke and forces it out with the return stroke, resulting in a pulsating delivery. Pumps of this type can have more than one cylinder, forming what is called a duplex (two-cylinder) or a triplex (three-cylinder) pump. Some reciprocating pumps draw water into the cylinder in front of the piston and discharge at the rear of the piston and are called double-acting pumps as opposed to the single-acting pumps where the water moves in one direction only with the movement of the piston. Although highly efficient and capable of increased capacity with increased engine speed these pumps have the disadvantage of being unable to handle water containing solids.

Displacement pumps of the diaphragm type can however handle liquids containing 10 to 15% of solids which makes them very popular. They work on the principle of raising and lowering a flexible diaphragm of rubber or rubberised canvas within a cylinder by means of a pump rod connected via a rocker bar to an engine crank. The upward movement of the diaphragm causes water to be sucked into the cylinder through a valve; the downward movement of the diaphragm closes the inlet valve and forces the water out through another valve into the delivery hose. Diaphragm or lift and force pumps are available with two cylinders and two diaphragms giving greater output and efficiency — typical pump examples are shown in Fig. VII.5.

Submersible pumps: used for extracting water from deep wells and sumps (see Fig. I.17) and are suspended in the water to be pumped. The power source is usually an electric motor to drive a centrifugal unit which is housed in a casing with an annular space to allow the water to rise upwards into the delivery pipe or rising main. Alternatively an electric submersible pump with a diaphragm arrangement can be used where large quantities of water are not involved.

ROLLERS

Rollers are designed to consolidate filling materials and to compact surface finishes such as tarmacadam for paths and pavings. The roller equipment used by building contractors is basically a smaller version of the large rollers used by civil engineering contractors for roadworks. Rollers generally rely upon deadweight to carry out the consolidating operation or by vibration as in the case of lightweight rollers. Deadweight rollers are usually diesel powered and driven by a seated operator within a cab. These machines can be obtained with weights ranging from 1 to 16 tonnes which is distributed to the ground through two large diameter rear wheels and a wider but small front steering-wheel. Many of these rollers carry water tanks to add to the dead load and to supply small sparge or sprinkler pipes fixed over the wheels to dampen the surfaces thus preventing the adhesion of tar or similar material when being rolled. These rollers are also available fitted with a scarifier to the rear of the vehicle for ripping up the surfaces of beds or roads — see Fig. VII.6.

Vibrating rollers which depend mainly upon the vibrations produced by the petrol- or diesel-powered engine can be hand guided or towed and are available with weights ranging from 500 kg to 5 tonnes — see Fig. VII.6. These machines will give the same degree of consolidation and compaction as their heavier deadweight counterparts, but being lighter and smaller they

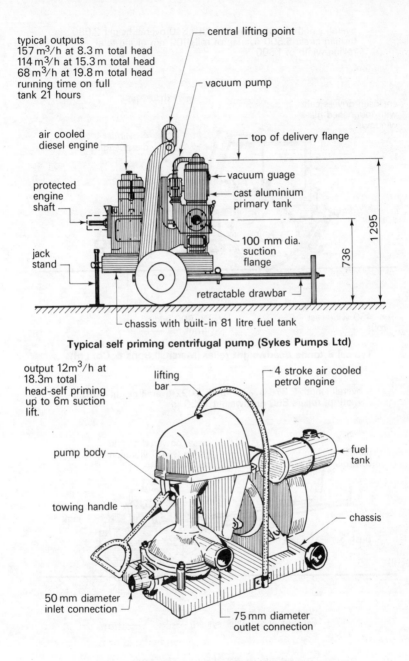

typical outputs
157 m³/h at 8.3 m total head
114 m³/h at 15.3 m total head
68 m³/h at 19.8 m total head
running time on full
tank 21 hours

central lifting point

vacuum pump

top of delivery flange

air cooled
diesel engine

protected
engine
shaft

vacuum guage

cast aluminium
primary tank

100 mm dia.
suction
flange

1295

736

jack
stand

retractable drawbar

chassis with built-in 81 litre fuel tank

Typical self priming centrifugal pump (Sykes Pumps Ltd)

output 12m³/h at
18.3m total
head-self priming
up to 6m suction
lift.

lifting
bar

4 stroke air cooled
petrol engine

pump body

fuel
tank

towing handle

chassis

50 mm diameter
inlet connection

75 mm diameter
outlet connection

Typical diaphragm pump (William R Selwood Ltd)

Fig VII.5 Typical pumps

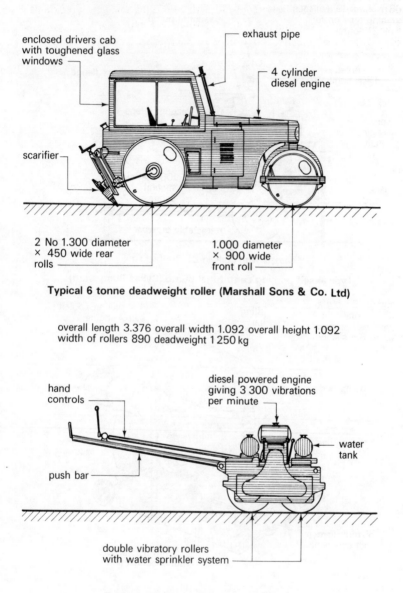

overall length 4.500 overall width 1.840 overall height 2.680
turning circle 5.500 overlap of rolls 100 mm giving total
rolling width of 1.600

enclosed drivers cab
with toughened glass
windows

exhaust pipe

4 cylinder
diesel engine

scarifier

2 No 1.300 diameter
× 450 wide rear
rolls

1.000 diameter
× 900 wide
front roll

Typical 6 tonne deadweight roller (Marshall Sons & Co. Ltd)

overall length 3.376 overall width 1.092 overall height 1.092
width of rollers 890 deadweight 1 250 kg

hand
controls

diesel powered engine
giving 3 300 vibrations
per minute

water
tank

push bar

double vibratory rollers
with water sprinkler system

Typical vibrating roller (Duomat R90)

Fig VII.6 Typical rollers

can be manoeuvred into buildings for consolidating small areas of hardcore or similar bed material. Single or double rollers are available with or without water sprinkler attachments and with vibrations within the region of 3 000 vibrations per minute. Vibrating rollers are particularly effective for the compaction and consolidation of granular soils.

16
Earth-moving and excavation plant

The selection, management and maintenance of builders' plant is particularly important when considered in the context of earth-moving and excavation plant. Before deciding to use any form of plant for these activities the site conditions and volume of work entailed must be such that it will be an economic venture. The difference between plant which is classified as earth-moving equipment and excavating machines is very narrow since a piece of plant which is designed primarily to excavate will also be capable of moving the spoil to an attendant transporting vehicle and likewise machines basically designed to move loose earth will also be capable of carrying out to some degree excavation works.

To browse through the catalogues of plant manufacturers and hirers to try and select a particular piece of plant is a bewildering exercise because of the wide variety of choice available for all classes of plant. Final choice is usually based upon experience, familiarity with a particular manufacturer's machines, availability or personal preference. There are many excellent works of reference devoted entirely to the analysis of the various machines to aid the would-be buyer or hirer; therefore in a text of this nature it is only necessary to consider the general classes of plant, pointing out their intended uses and amplifying this with typical examples of the various types without claiming that the example chosen is the best of its type but only representative.

BULLDOZERS AND ANGLEDOZERS
These machines are primarily a high-powered tractor with caterpillar or crawler tracks fitted with a mould board or

blade at the front for stripping and oversite excavations up to a depth of 400 mm (depending upon machine specification) by pushing the loosened material ahead of the machine. For back-filling operations the angledozer with its mould board set at an angle, in plan, to the machine's centre line can be used. Most mould boards can be set at an angle either in the vertical or horizontal plane to act as an angledozer and on some models the leading edge of the mould board can be fitted with teeth for excavating in hard ground. These machines can be very large with mould boards of 1.200 to 4.000 in width x 600 mm to 1.200 in height with a depth of cut up to 400 mm. Most bulldozers and angledozers are mounted on crawler tracks, although small bulldozers with a wheeled base are available. The control of the mould board on most models is hydraulic, the alternative being a winch and wire cable control system — for typical example see Fig. VII.7. In common with other tracked machines one of the disadvantages of this arrangement is the need for a special transporting vehicle such as a low loader to move the equipment between sites.

Before any earth-moving work is started a drawing should be produced indicating the areas and volumes of cut and fill required to enable a programme to be prepared to reduce machine movements to a minimum. When large quantities of earth have to be moved on a cut and fill basis to form a predetermined level or gradient it is good practice to draw up a mass haul diagram indicating the volumes of earth to be moved, the direction of movement and the need to import more spoil or alternatively cart away the surplus — see Fig. VII.8 for typical example.

SCRAPERS

This piece of plant consists of a power unit and a scraper bowl and is used to excavate and transport soil where surface stripping, site levelling and cut and fill activities are planned, particularly where large volumes are involved. These machines are capable of producing a very smooth and accurate formation level and come in three basic types:

1. Crawler-drawn scraper.
2. Two-axle scraper.
3. Three-axle scraper.

The design and basic operation of the scraper bowl is similar in all three types, consisting of a shaped bowl with a cutting edge which can be lowered to cut the top surface of the soil up to a depth of 300 mm. As the bowl moves forward the loosened earth is forced into the container and when full the cutting edge is raised to seal the bowl. To ensure that a full load is obtained many contractors use a bulldozer to act as a pusher over

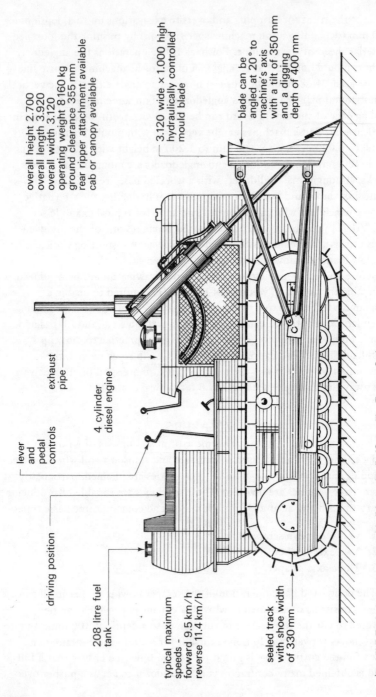

overall height 2.700
overall length 3.920
overall width 3.120
operating weight 8 160 kg
ground clearance 355 mm
rear ripper attachment available
cab or canopy available

3.120 wide × 1.000 high
hydraulically controlled
mould blade

blade can be
angled at 20° to
machine's axis
with a tilt of 350 mm
and a digging
depth of 400 mm

exhaust
pipe

lever
and
pedal
controls

4 cylinder
diesel engine

driving position

208 litre fuel
tank

typical maximum
speeds –
forward 9.5 km/h
reverse 11.4 km/h

sealed track
with shoe width
of 330 mm

Fig VII.7 Typical tractor powered bulldozer details (Caterpiller Tractor Co.)

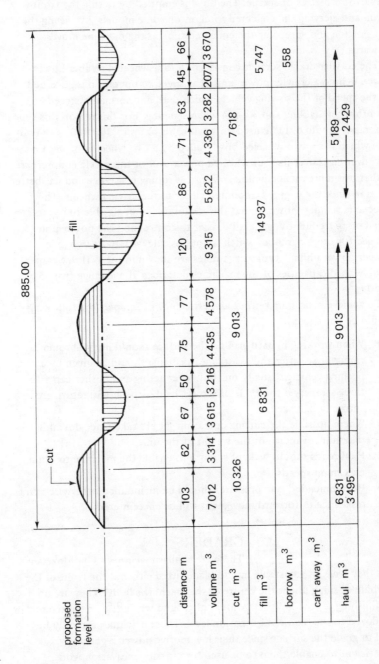

distance m	103	62	67	50	75	77	120	86	71	63	45	66
volume m³	7 012	3 314	3 615	3 216	4 435	4 578	9 315	5 622	4 336	3 282	2 077	3 670
cut m³	10 326		9 013		9 013				7 618			
fill m³			6 831				14 937				5 747	
borrow m³											558	
cart away m³												
haul m³	6 831 / 3 495			9 013					5 189 / 2 429			

Fig VII.8 Typical mass haul diagram

885.00

cut

fill

proposed
formation
level

the last few metres of scrape. The bowl is emptied by raising the front apron and ejecting the collected spoil or, on some models, by raising the rear portion and spreading the collected spoil as the machine moves forwards.

The crawler-drawn scraper consists of a four-wheeled scraper bowl towed behind a crawler power unit. The speed of operation is governed by the speed of the towing vehicle which does not normally exceed 8 km/h when hauling and 3 km/h when scraping. For this reason this type of scraper should only be used on small hauls of up to 300.000. The two-axle which has a two-wheeled bowl pulled by a two-wheeled power unit has advantages over its four-wheeled power unit or three-axle counterpart in that it is more manoeuvrable, offers less rolling resistance and has better traction since the engine is mounted closer to the driving wheels. The three-axle scraper, however, has the advantages of being able to use its top speed more frequently, generally easier to control and the power unit can be used for other activities which is not possible with most two-axle scraper power units — typical examples are shown in Fig. VII.9. Scraper bowl heaped capacities of the machines described above range from 5 to 50 m^3.

To achieve maximum output and efficiency of scrapers the following should be considered:

1. When working in hard ground the surface should be pre-broken by a ripper or scarifier and assistance in cutting should be given by a pushing vehicle. Usually one bulldozer acting as a pusher can assist three scrapers if the cycle of scrape, haul, deposit and return are correctly balanced.
2. Where possible the cutting operation should take place downhill to take full advantage of the weight of the unit.
3. Haul roads should be kept smooth to enable the machine to obtain maximum speeds.
4. Recommended tyre pressures should be maintained otherwise extra resistance to forward movement will be encountered.

GRADERS

These are similar machines to bulldozers in that they have an adjustable mould blade fitted either at the front of the machine or slung under the centre of the machine's body. They are used for finishing to fine limits large areas of ground which have been scraped or bulldozed to the required formation level. These machines can only be used to grade the surface since their low motive power is generally insufficient to enable them to be used for oversite excavation work.

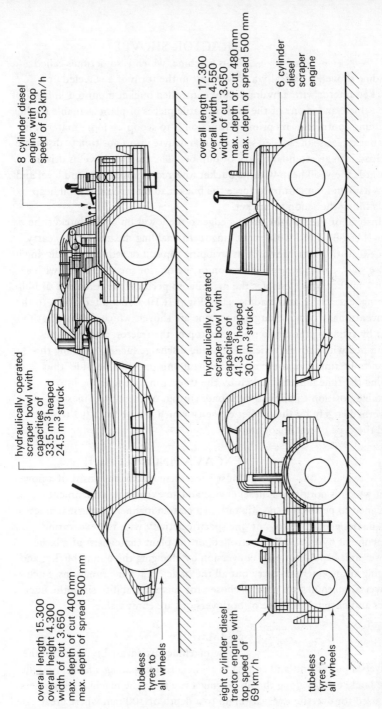

8 cylinder diesel engine with top speed of 53 km/h

hydraulically operated scraper bowl with capacities of 33.5 m³ heaped 24.5 m³ struck

overall length 15.300
overall height 4.300
width of cut 3.650
max. depth of cut 400 mm
max. depth of spread 500 mm

tubeless tyres to all wheels

overall length 17.300
overall width 4.550
width of cut 3.650
max. depth of cut 480 mm
max. depth of spread 500 mm

6 cylinder diesel scraper engine

hydraulically operated scraper bowl with capacities of 41.3 m³ heaped 30.6 m³ struck

eight cylinder diesel tractor engine with top speed of 69 km/h

tubeless tyres to all wheels

Fig VII.9 Typical 2 and 3 axle scraper details (Caterpiller Tractor Co.)

209

TRACTOR SHOVEL

This machine, which is sometimes called a loading shovel, is basically a power unit in the form of a wheeled or tracked tractor with a hydraulically controlled bucket mounted in front of the vehicle and is one of the most versatile pieces of plant available to the building contractor. Its primary function is to scoop up loose material in the bucket, raise the loaded spoil and manoeuvre into a position to discharge its load into an attendant lorry or dumper. The tractor shovel is driven towards the spoil heap with its bucket lowered almost to ground level and uses its own momentum to force the bucket to bite into the spoil heap thus filling the scoop or bucket.

Instead of the straight cutting edge to the lower lip of the bucket the shovel can be fitted with excavating teeth enabling the machine to carry out excavating activities such as stripping top soil or reduce level digging in loose soils. Another popular version of the tractor shovel is fitted with a 4-in-1 bucket which enables the machine to perform the functions of bulldozing, excavating and loading — see Fig. VII.10. Other alternatives to the conventional front-discharging machine are shovels which discharge at the rear by swinging the bucket over the top of the tractor, and machines equipped with shovels which have a side discharge facility enabling the spoil to be tipped into the attendant haul unit parked alongside, thus saving the time normally taken by the tractor in manoeuvring into a suitable position to discharge its load. Output of these machines is governed largely by the bucket capacity which can be from 0.5 to 4 m^3 and the type of soil encountered.

EXCAVATING MACHINES

Most excavating machines consist of a power unit which is normally a diesel engine and an excavating attachment designed to perform a specific task in a certain manner. These machines can be designed to carry out one specific activity with the excavating attachment hydraulically controlled, or the plant can consist of a basic power unit capable of easy conversion by changing the boom, bucket and rigging arrangement to carry out all the basic excavating functions. Such universal machines are usually chosen for this adaptability since the bucket sizes and outputs available of both versions are comparable.

SKIMMER

These machines are invariably based on the universal power unit and consist of a bucket sliding along a horizontal jib. The bucket slides along the jib digging away from the machine. Skimmers are used for oversite excavation up to a depth of 300 mm where great

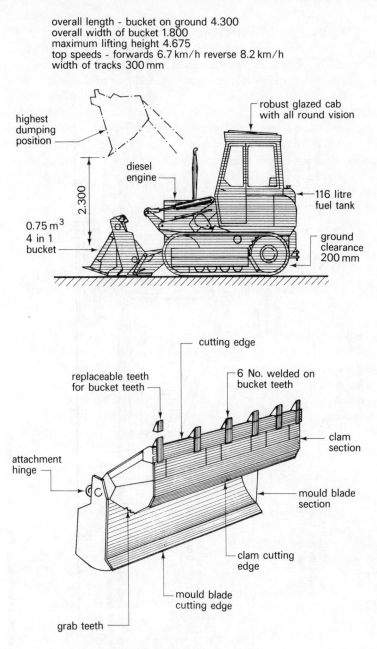

overall length - bucket on ground 4.300
overall width of bucket 1.800
maximum lifting height 4.675
top speeds - forwards 6.7 km/h reverse 8.2 km/h
width of tracks 300 mm

highest dumping position

robust glazed cab with all round vision

diesel engine

116 litre fuel tank

2.300

0.75 m³ 4 in 1 bucket

ground clearance 200 mm

cutting edge

replaceable teeth for bucket teeth

6 No. welded on bucket teeth

attachment hinge

clam section

mould blade section

clam cutting edge

mould blade cutting edge

grab teeth

Typical 4 in 1 bucket details

Fig VII.10 Typical tractor shovel (International Harvester Co)

212

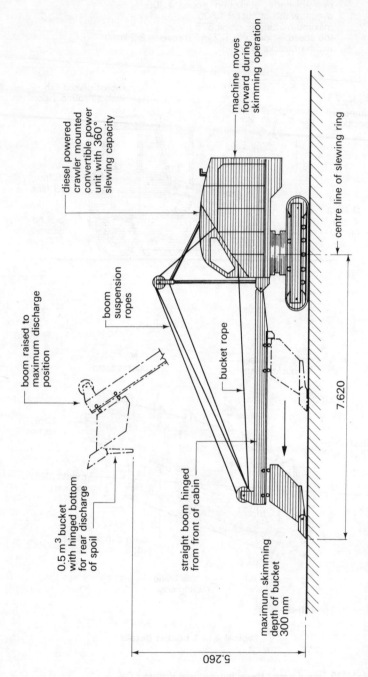

boom raised to maximum discharge position

0.5 m³ bucket with hinged bottom for rear discharge of spoil

boom suspension ropes

bucket rope

diesel powered crawler mounted convertible power unit with 360° slewing capacity

machine moves forward during skimming operation

centre line of slewing ring

straight boom hinged from front of cabin

maximum skimming depth of bucket 300 mm

7.620

5.260

Fig VII.11 Typical skimmer details (Ruston-Bucyrus Limited)

accuracy in level is required and they can achieve an output of some 50 bucket loads per hour. To discharge the spoil the boom or jib is raised and the power unit is rotated until the raised bucket is over the attendant haulage vehicle, enabling the spoil to be discharged through the opening bottom direct into the haul unit — see Fig. VII.11.

FACE SHOVEL

This type of machine can be used as a loading shovel or for excavating into the face of an embankment or berm. Universal power unit or hydraulic machines are available with a wide choice of bucket capacities achieving outputs in the region of 80 bucket loads per hour. The discharge operation is similar to that described above for the skimmer except that in the universal machine the discharge opening is at the rear of the bucket whereas in the hydraulic machines discharge is from the front of the bucket — see Figs. VII.10 and 12. These machines are limited in the depth to which they can dig below machine level; this is generally within the range of 300 mm to 2.000.

BACKACTOR

This piece of plant is probably the most common form of excavating machinery used by building contractors for excavating basements, pits and trenches. Universal power unit and hydraulic versions are available, the latter often sacrificing bucket capacity to achieve a greater reach from a set position. Discharge in both types is by raising the bucket in a tucked position and emptying the spoil through the open front end into the attendant haul unit or alongside the trench. Outputs will vary from 30 to 60 bucket loads per hour, depending upon how confined is the excavation area. Typical details are shown in Figs. VII.13 and 14.

DRAGLINE

This type of excavator is essentially a crane with a long jib to which is attached a drag bucket for excavating in loose and soft soils below the level of the machine. This machine is for bulk excavation where fine limits are not of paramount importance since this is beyond the capabilities of the machine's design. The accuracy to which a dragline can excavate depends upon the skill of the operator. Discharge of the collected spoil is similar to that of a backactor, being through the open front end of the bucket — see Fig. VII.15. A machine rigged as a dragline can be fitted with a grab bucket as an alternative for excavating in very loose soils below the level of the machine. Outputs of dragline excavators will vary according to operating restrictions from 30 to 80 bucket loads per hour.

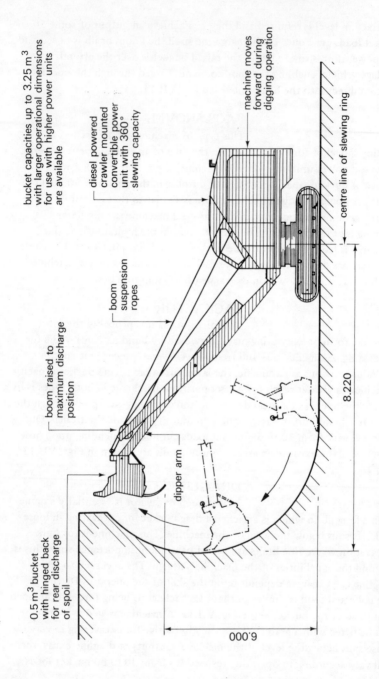

bucket capacities up to 3.25 m³ with larger operational dimensions for use with higher power units are available

diesel powered crawler mounted convertible power unit with 360° slewing capacity

machine moves forward during digging operation

centre line of slewing ring

boom suspension ropes

boom raised to maximum discharge position

dipper arm

0.5 m³ bucket with hinged back for rear discharge of spoil

8.220

6.000

Fig VII.12 Typical face shovel details (Ruston-Bucyrus Limited)

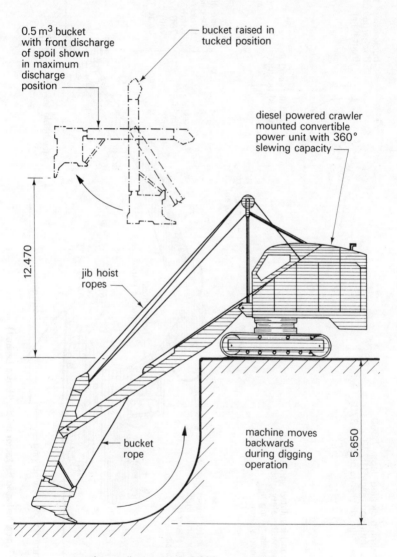

bucket capacities up to 1.53 m³
with larger operational dimensions
for use with larger power machines
are available

0.5 m³ bucket
with front discharge
of spoil shown
in maximum
discharge
position

bucket raised in
tucked position

diesel powered crawler
mounted convertible
power unit with 360°
slewing capacity

12.470

jib hoist
ropes

bucket
rope

machine moves
backwards
during digging
operation

5.650

maximum digging reach 9.600 measured from cutting
edge of bucket to centre line of slewing ring

Fig VII.13 Typical backactor details (Ruston-Bucyrus Ltd)

215

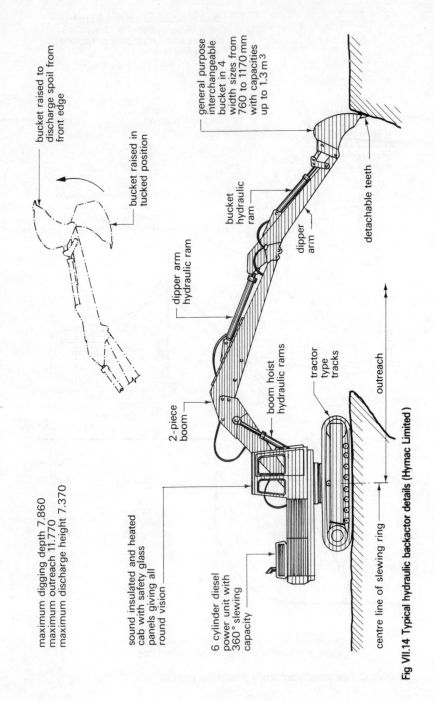

bucket raised to discharge spoil from front edge

bucket raised in tucked position

maximum digging depth 7.860
maximum outreach 11.770
maximum discharge height 7.370

general purpose interchangeable bucket in 4 width sizes from 760 to 1170 mm with capacities up to 1.3 m³

detachable teeth

bucket hydraulic ram

dipper arm

dipper arm hydraulic ram

tractor type tracks

2-piece boom

boom hoist hydraulic rams

outreach

centre line of slewing ring

sound insulated and heated cab with safety glass panels giving all round vision

6 cylinder diesel power unit with 360° slewing capacity

Fig VII.14 Typical hydraulic backactor details (Hymac Limited)

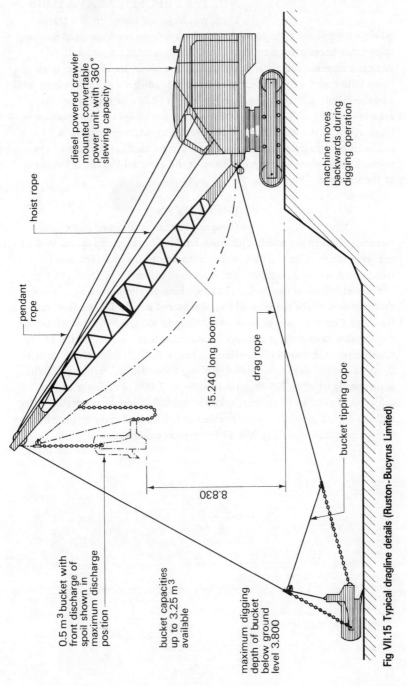

diesel powered crawler mounted convertable power unit with 360° slewing capacity

machine moves backwards during digging operation

hoist rope

pendant rope

drag rope

bucket tipping rope

15.240 long boom

8.830

0.5 m³ bucket with front discharge of spoil shown in maximum discharge position

bucket capacities up to 3.25 m³ available

maximum digging depth of bucket below ground level 3.800

Fig VII.15 Typical dragline details (Ruston-Bucyrus Limited)

217

MULTI-PURPOSE EXCAVATORS

These machines are based upon a tractor power unit and are very popular with the small- to medium-sized building contractor because of their versatility. The tractor is usually a diesel-powered wheeled vehicle although tracked versions are available, both being fitted with a hydraulically controlled loading shovel at the front and a hydraulically controlled backacting bucket or hoe at the rear of the vehicle — see Fig. VII.16. It is essential that the weight of the machine is removed from the axles during a backacting excavation operation. This is achieved by outrigger jacks at the corners or by jacks at the rear of the power unit working in conjunction with the inverted bucket at the front of the machine.

TRENCHERS

These are machines designed to excavate trenches of constant width with considerable accuracy and speed. Widths available range from 250 to 450 mm with depths up to 4.000. Most trenchers work on a conveyor principle having a series of small cutting buckets attached to two endless chains which are supported by a boom that is lowered into the ground to the required depth. The spoil is transferred to a cross conveyor to deposit the spoil alongside the trench being dug, or alternatively it is deposited onto plough-shaped deflection plates which direct the spoil into continuous heaps on both sides of the trench being excavated as the machine digs along the proposed trench run. With a depth of dig of some 1.500 outputs of up to 2.000 per minute can be achieved, according to the nature of the subsoil. Some trenchers are fitted with an angled mould blade to enable the machine to carry out the back-filling operation — see Fig. VII.17 for typical example.

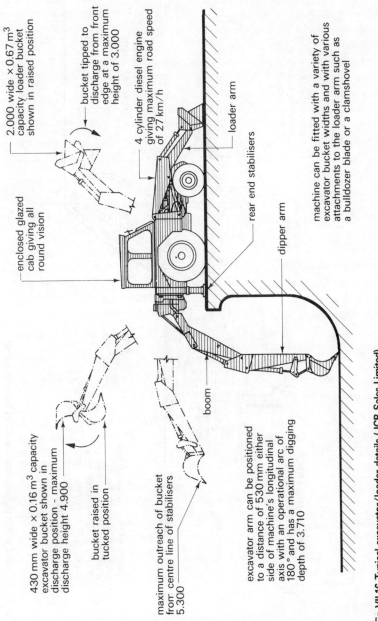

2.000 wide ×0.67 m³ capacity loader bucket shown in raised position

bucket tipped to discharge from front edge at a maximum height of 3.000

4 cylinder diesel engine giving maximum road speed of 27 km/h

loader arm

rear end stabilisers

dipper arm

machine can be fitted with a variety of excavator bucket widths and with various attachments to the loader arm such as a bulldozer blade or a clamshovel

enclosed glazed cab giving all round vision

boom

430 mm wide × 0.16 m³ capacity excavator bucket shown in discharge position - maximum discharge height 4.900

bucket raised in tucked position

maximum outreach of bucket from centre line of stabilisers 5.300

excavator arm can be positioned to a distance of 530 mm either side of machine's longitudinal axis with an operational arc of 180° and has a maximum digging depth of 3.710

Fig VII.16 Typical excavator/loader details (JCB Sales Limited)

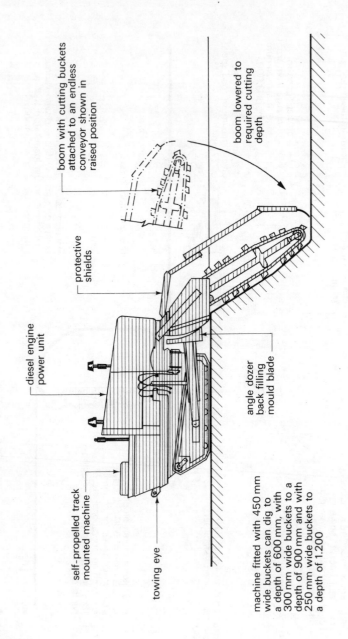

boom with cutting buckets attached to an endless conveyor shown in raised position

boom lowered to required cutting depth

protective shields

diesel engine power unit

angle dozer back filling mould blade

self-propelled track mounted machine

towing eye

machine fitted with 450 mm wide buckets can dig to a depth of 600 mm, with 300 mm wide buckets to a depth of 900 mm and with 250 mm wide buckets to a depth of 1.200

Fig VII.17 Typical trench digging machine (Davis Manufacturing)

220

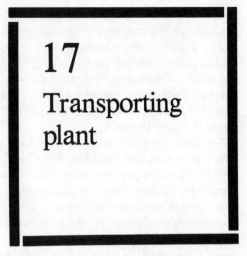

17
Transporting plant

Movement of materials and/or personnel around and between building sites can be very time-consuming and non-productive; therefore wherever economically possible most contractors will use some form of mechanical transportation. The movement required can be horizontal, vertical or a combination of both directions. In the case of horizontal and vertical movement of large quantities of water the usual plant employed is a pump (already described in the chapter on small powered plant). Similarly the transportation of large quantities of concrete can be carried out by using special pumping equipment and this form of material transportation will be considered in the chapter on concreting plant.

LORRIES AND TRUCKS

Transportation between sites of men, machines and materials is usually carried out by using suitably equipped or adapted lorries or trucks ranging from the small 'pick-up' vehicle weighing less than 800 kg unladen to the very long, low loaders used to convey tracked plant such as cranes and bulldozers. The small 'pick-up' vehicle is usually based upon the manufacturer's private car range and has the advantages of lower road tax than heavier lorries and that the driver does not require a heavy goods vehicle licence for trucks weighing less than 3 000 kg unladen. Most lorries designed and developed for building contractors' use are powered by a diesel, which is more economical to operate than the petrol engine although they are heavier and dearer. But with mileages in excess of 32 000 km per year experienced by most contractors diesel engines usually prove to be a worth-while proposition. A vast range of lorries are produced by the leading motor manufacturers and are available with refinements such as tipping, tailhoist and self-loading facilities using hydraulic lifting gear. Since lorries, trucks and vans are standard forms of transportation encountered in all aspects of daily living they will not be considered in detail in this text but regulations made under the Road Traffic Act, Vehicles (Excise) Act and the Customs and Excise Act must be noted.

The above legislation is very extensive and complex, dealing in detail with such requirements as the minimum driving ages for various types of vehicles; limitations of hours of duty; maximum speed limits of certain classes of vehicles; vehicle lighting regulations; construction, weight and equipment of motor vehicles and trailers; testing requirements and the rear-marking regulations for vehicles over 13.000 long. Most of the statutory requirements noted above will be incorporated into the design and finish of the vehicle as purchased, but certain regulations regarding such matters as projection of loads, maximum loading of vehicles and notifications to be given to highway authorities, police forces and the Ministry of Transport are the direct concern of the building contractor. Precise information on these requirements can be found in the Motor Vehicles (Authorisation of Special Types) General Order 1969 together with any subsequent amendments. It should be noted that excise regulations and the use of rebated fuels are also the builder's responsibility.

Dumpers

These are one of the most versatile, labour-saving and misused pieces of plant available to the builder for the horizontal movement of materials ranging from bricks to aggregates, sanitary fittings to scaffolding and fluids such as wet concrete. These diesel-powered vehicles require only one operative, the driver, and can transverse the rough terrain encountered on many building sites. Many sizes and varieties are produced, giving options such as two- or four-wheel drive, hydraulic- or gravity-operated container, side or high level discharge, self-loading facilities and specially equipped dumpers for collecting and transporting crane skips. Specification for dumpers is usually given by quoting the container capacity in litres for heaped, struck and water levels — see Fig. VII.18 for typical examples.

Fork lift trucks

Fork lift trucks for handling mainly palleted materials quickly and efficiently around building sites over the rough terrain normally encountered have rapidly gained in popularity since their

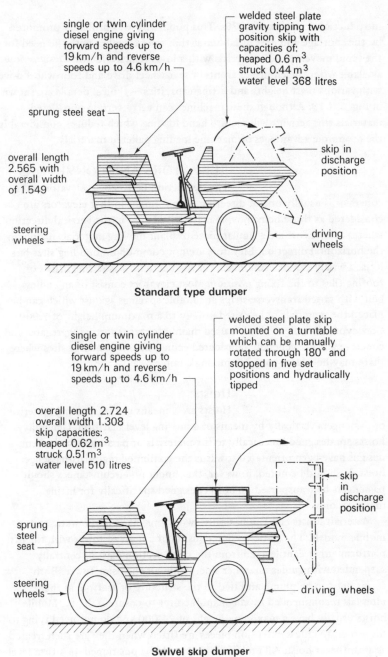

single or twin cylinder diesel engine giving forward speeds up to 19 km/h and reverse speeds up to 4.6 km/h

welded steel plate gravity tipping two position skip with capacities of: heaped 0.6 m³ struck 0.44 m³ water level 368 litres

sprung steel seat

overall length 2.565 with overall width of 1.549

skip in discharge position

steering wheels

driving wheels

Standard type dumper

single or twin cylinder diesel engine giving forward speeds up to 19 km/h and reverse speeds up to 4.6 km/h

welded steel plate skip mounted on a turntable which can be manually rotated through 180° and stopped in five set positions and hydraulically tipped

overall length 2.724 overall width 1.308 skip capacities: heaped 0.62 m³ struck 0.51 m³ water level 510 litres

skip in discharge position

sprung steel seat

steering wheels

driving wheels

Swivel skip dumper

Fig VII.18 Typical diesel dumpers (Liner Concrete Machinery Co Ltd)

introduction in the early 1970s. This popularity was probably promoted by the shortage and cost of labour at that time together with the need for the rapid movement of materials with a low breakage factor. Designs now available offer the choice of front- or rear-wheel drive and four-wheel drive with various mast heights and lifting capacities — typical details are shown in Fig. VII.19. Although these machines can carry certain unpalleted materials this activity will require hand loading which reduces considerably the economic advantages of machine loading palleted materials.

ELEVATORS AND CONVEYORS

The distinction between an elevator and a conveyor is usually one of direction of movement in that elevators are considered as those belts moving materials mainly in the vertical direction whereas conveyors are a similar piece of plant moving materials mainly in the horizontal direction. Elevators are not common on building sites but if used wisely can be economical for such activities as raising bricks or roofing tiles to the fixing position. Most elevators consist of an endless belt with raised transverse strips at suitable spacings against which can be placed the materials to be raised usually to a maximum height of 7.000. Conveyors or endless belts are used mainly for transporting aggregates and concrete and are generally considered economic only on large sites where there may be a large concrete-mixing complex.

Hoists

Hoists are a means of transporting materials or passengers vertically by means of a moving level platform. Generally, hoists are designed specifically to lift materials or passengers but recent designs have been orientated towards the combined materials/passenger hoist. It should be noted, however, that under no circumstances should passengers be transported on hoists designed specifically for lifting materials only.

Materials hoists come in basically two forms, namely the static and mobile models. The static version consists of a mast or tower with the lift platform either cantilevered from the small section mast or centrally suspended with guides on either side within an enclosing tower. Both forms need to be plumb and tied to the structure or scaffold at the intervals recommended by the manufacturer to ensure stability. Mobile hoists usually have a maximum height of 24.000 and do not need tying to the structure unless extension pieces are fitted when they are then treated as a cantilever hoist. All mobile hoists should be positioned on a firm level base and jacked to ensure stability — see Fig. VII.20. The operation of a materials hoist should be entrusted to a trained driver who has a clear view

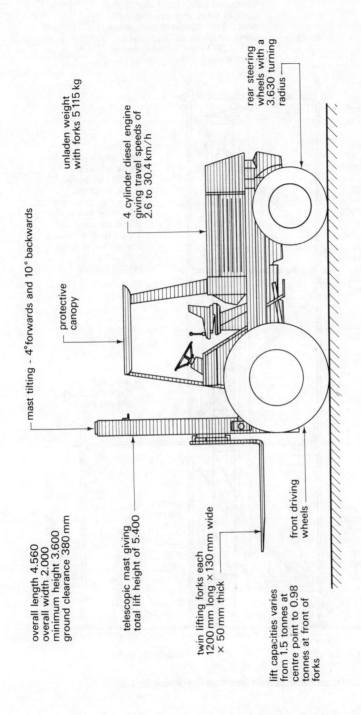

overall length 4.560
overall width 2.000
minimum height 3.600
ground clearance 380 mm

mast tilting - 4° forwards and 10° backwards

unladen weight
with forks 5 115 kg

protective
canopy

4 cylinder diesel engine
giving travel speeds of
2.6 to 30.4 km/h

rear steering
wheels with a
3.630 turning
radius

telescopic mast giving
total lift height of 5.400

twin lifting forks each
1200 mm long × 130 mm wide
× 50 mm thick

lift capacities varies
from 1.5 tonnes at
centre point to 0.98
tonnes at front of
forks

front driving
wheels

Fig VII.19 Typical fork lift truck details (Manitou (Site Lift) Ltd)

225

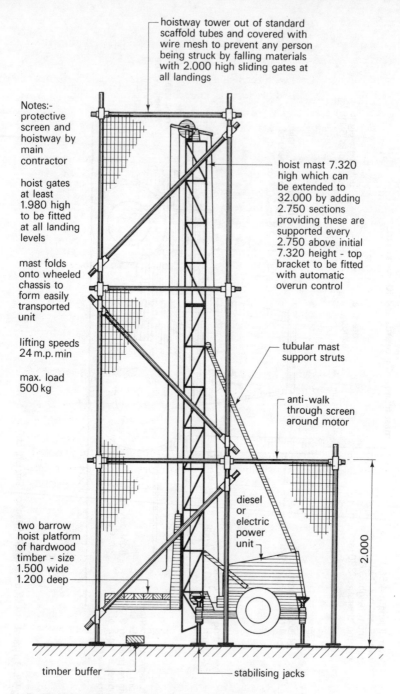

hoistway tower out of standard scaffold tubes and covered with wire mesh to prevent any person being struck by falling materials with 2.000 high sliding gates at all landings

Notes:-
protective screen and hoistway by main contractor

hoist gates at least 1.980 high to be fitted at all landing levels

mast folds onto wheeled chassis to form easily transported unit

lifting speeds 24 m.p. min

max. load 500 kg

two barrow hoist platform of hardwood timber - size 1.500 wide 1.200 deep

hoist mast 7.320 high which can be extended to 32.000 by adding 2.750 sections providing these are supported every 2.750 above initial 7.320 height - top bracket to be fitted with automatic overrun control

tubular mast support struts

anti-walk through screen around motor

diesel or electric power unit

2.000

timber buffer

stabilising jacks

Fig VII.20 Typical materials hoist (Wickham Engineering Co. Ltd)

from the operating position. Site operatives should be instructed as to the correct loading procedures, such as placing barrows onto the hoist platform at ground level with the handles facing the high level exit so that walking onto the raised platform is reduced to a minimum.

Passenger hoists, like the materials hoist, can be driven by a petrol, diesel or electric motor and can be of a cantilever or enclosed variety. The cantilever type consists of one or two passenger hoist cages operating on one or both sides of the cantilever tower; the alternative form consists of a passenger hoist cage operating within an enclosing tower. Tying back requirements are similar to those needed for materials hoist. Passenger hoists should conform to the recommendations of BS 4465 and materials hoists with the recommendations of BS 3125. Typical hoist details are shown in Fig. VII.21.

THE CONSTRUCTION (LIFTING OPERATIONS) REGULATIONS 1961

This Statutory Instrument sets out in Parts V and VI the legal requirements regarding the use of hoists on construction sites.

Reg. 42: gives details regarding the need to enclose the hoistway wherever access can be gained and wherever anyone at ground level could be struck by the platform or counterweight and such enclosures and gates should be at least 2.000 high. Access gates must be kept closed at all times except for the necessary loading and unloading of the platform. The platform itself must be fitted with a device capable of supporting a fully loaded platform in the event of failure of the hoist ropes or hoisting gear; furthermore the hoist must be fitted with an automatic device to prevent the platform or cage over-running its highest point.

Reg. 43: deals with the operation of hoists, requiring that the controlling of the hoist is to be from one position only at all times if not controlled from the cage itself and that the driver must have a clear view of the hoist throughout its entire travel; if this is not possible a signalling system covering all landings must be installed and used.

Reg. 44: sets out the requirements for winches which must have an automatic braking system which is applied whenever the control lever, handle or switch is not in the operating position.

Reg. 45: deals with safe working loads in terms of materials and/or maximum number of passengers to be carried, and such safe working loads must be displayed on all platforms or cages.

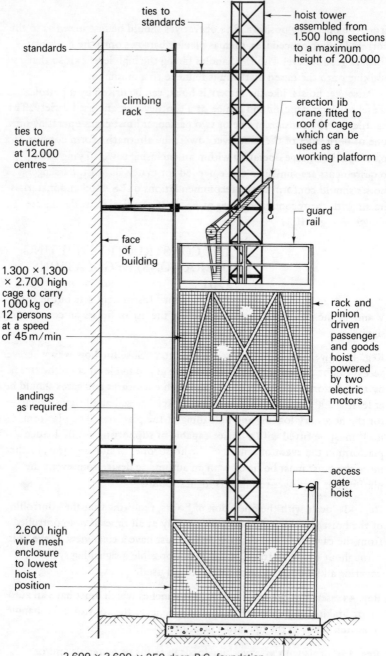

ties to standards

standards

hoist tower assembled from 1.500 long sections to a maximum height of 200.000

climbing rack

ties to structure at 12.000 centres

erection jib crane fitted to roof of cage which can be used as a working platform

face of building

guard rail

1.300 × 1.300 × 2.700 high cage to carry 1 000 kg or 12 persons at a speed of 45 m/min

rack and pinion driven passenger and goods hoist powered by two electric motors

landings as required

access gate hoist

2.600 high wire mesh enclosure to lowest hoist position

2.600 × 3.600 × 250 deep R.C. foundation

Fig VII.21 Typical passenger hoist (Linden Alimark Ltd)

228

Reg. 46: sets out the testing and examination requirements for hoists which are:

1. Testing and thorough examination — before being used for the first time, after alterations to height and after repairs.
2. Thorough examination — at least every six months.
3. Inspection — weekly.

All tests, examinations and inspections are to be carried out by a competent person and recorded on the appropriate form 91 (Part 1).

Reg. 47: covers lifting devices which may be used to carry passengers. These include power-driven hoists, power-driven suspended scaffolds and, provided certain precautions are taken to stop passengers falling out, buckets or skips attached to a crane.

Reg. 48: deals specifically with passenger-carrying hoists and requires the cages to be constructed in such a manner that passengers cannot fall out, become trapped or struck by objects falling down the hoistway. Other requirements under this regulation include the need for gates which will prevent the hoists being activated until they are closed and which can only be opened at landing levels. Overrun devices must also be fitted at the bottom of the hoistway as well as those at the top in accordance with Regulation 42.

Reg. 49: is concerned with the security of loads being transported by a hoist, such as loose materials, which should be lifted in suitable containers, and wheelbarrows, which should be scotched or otherwise suitably secured to prevent movement or tipping over during transportation.

CRANES

A crane may be defined as a device or machine for lifting loads by means of a rope. The use of cranes has greatly increased in the construction industry due mainly to the need to raise the large and heavy prefabricated components often used in modern structures. The range of cranes available is very wide and therefore actual choice must be made on a basis of sound reasoning, overall economics, capabilities of cranes under consideration, prevailing site conditions and the anticipated utilisation of the equipment.

The simplest crane of all consists of a single-grooved wheel, over which the rope is passed, suspended from a scaffold or beam and is called a gin wheel. The gin wheel is manually operated and always requires more effort than the weight of the load to raise it to the required height. It is only suitable for light loads as, for example, a bucketful of mortar and is normally only used on very small contracts. To obtain some mechanical advantage the gin wheel can be replaced by a pulley block which contains more than one pulley or sheave; according to the number and pattern of sheaves used the lesser or greater is the saving in effort required to move any given load.

Another useful but simple crane which can be employed for small, low rise structures is the scaffold crane which consists of a short jib counterbalanced by the small petrol or electric power unit. The crane is fastened to a specially reinforced scaffold standard incorporated within the general scaffold framework with extra bracing to overcome the additional stresses as necessary. The usual maximum lifting capacity of this form of crane is 200 kg.

Apart from these simple cranes for small loads most cranes come in the more recognisable form. Subdivision of crane types can be very wide and varied but one simple method of classification is to consider cranes under three general headings:

1. Mobile cranes.
2. Static or stationary cranes.
3. Tower cranes.

Mobile cranes

Mobile cranes come in a wide variety of designs and capacities, generally with a 360° rotation or slewing circle, a low pivot and luffing jib, the main exception being the mast crane. Mobile cranes can be classed into five groups:

1. Self-propelled cranes.
2. Lorry-mounted cranes.
3. Track-mounted cranes.
4. Mast cranes.
5. Gantry cranes.

Self-propelled cranes: these are wheel-mounted mobile cranes which are generally of low lifting capacities of up to 10 tonnes. They can be distinguished from other mobile cranes by the fact that the driver has only one cab position for both driving and operating the crane. They are extremely mobile but to be efficient they usually require a hard level surface from which to work. Road speeds obtained are in the region of 30 km/h. The small capacity machines have a fixed boom or jib length, whereas the high capacity cranes can have a sectional lattice jib or a telescopic boom to obtain various radii and lifting capacities. In common with all cranes the shorter the lifting radius the greater will be the lifting capacity — see Fig. VII.22 for typical example.

230

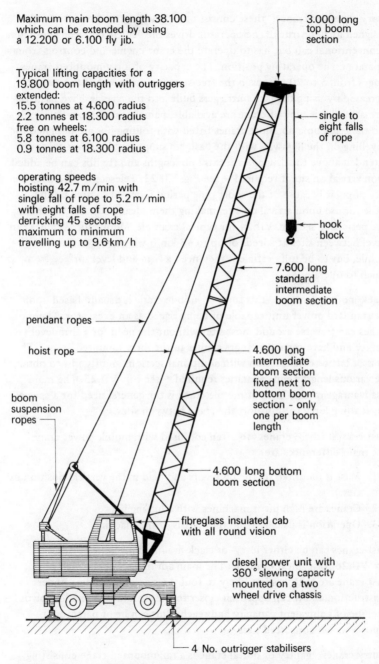

Maximum main boom length 38.100
which can be extended by using
a 12.200 or 6.100 fly jib.

Typical lifting capacities for a
19.800 boom length with outriggers
extended:
15.5 tonnes at 4.600 radius
2.2 tonnes at 18.300 radius
free on wheels:
5.8 tonnes at 6.100 radius
0.9 tonnes at 18.300 radius

operating speeds
hoisting 42.7 m/min with
single fall of rope to 5.2 m/min
with eight falls of rope
derricking 45 seconds
maximum to minimum
travelling up to 9.6 km/h

3.000 long
top boom
section

single to
eight falls
of rope

hook
block

7.600 long
standard
intermediate
boom section

pendant ropes

hoist rope

4.600 long
intermediate
boom section
fixed next to
bottom boom
section - only
one per boom
length

boom
suspension
ropes

4.600 long bottom
boom section

fibreglass insulated cab
with all round vision

diesel power unit with
360° slewing capacity
mounted on a two
wheel drive chassis

4 No. outrigger stabilisers

Fig VII.22 Typical self propelled crane (Jones Cranes Ltd)

231

Lorry-mounted cranes: these consist of a crane mounted on a specially designed lorry or truck. The operator drives the vehicle between sites from a conventional cab but has to operate the crane engine and controls from a separate crane operating position. The capacity of lorry-mounted cranes ranges from 5 to 20 tonnes in the free-standing position but this can be increased by using the jack outriggers built into the chassis. Two basic jib formats for this type of crane are available, namely the folding lattice jib and the telescopic jib. Most cranes fitted with folding jibs are designed for travelling on the highway with the basic jib supported by a vertical frame extended above the driving cab; extra jib lengths and fly jibs can be added upon arrival on site if required — see Fig. VII.23. Telescopic jib cranes are very popular because of the short time period required to prepare the crane for use upon arrival on site, making them ideally suitable for short-hire periods — see Fig. VII.24 for typical example. Mobile lorry cranes can travel between sites at speeds of up to 48 km/h which makes them very mobile, but to be fully efficient they need a firm and level surface from which to operate.

Track-mounted cranes: this form of mobile crane is usually based upon the standard power unit capable of being rigged as an excavator. These cranes can traverse around most sites without the need for a firm level surface and have capacity ranges similar to the lorry-mounted cranes. The jib is of lattice construction with additional sections and fly jibs to obtain the various lengths and capacities required — see Fig. VII.25. The main disadvantage of this form of mobile crane is the general need for a special low-loading lorry to transport the crane between sites.

Mast cranes: these cranes are often confused with mobile tower cranes. The main differences are:

1. Mast is mounted on the jib pivots and held in the vertical position by ties.
2. Cranes are high pivot machines with a luffing jib.
3. Operation is usually from the chassis of the machine.

Mast cranes can be either lorry- or track-mounted machines — see Fig. VII.26 for typical example. The main advantages of the high pivot mast crane are that it is less likely to foul the side of a building under construction and it can approach closer to the structure than a low pivot machine of equivalent capacity and reach. This can be of paramount importance on congested sites.

Gantry cranes: gantry or portal crane is a rail-mounted crane consisting of a horizontal transverse beam which carries a combined driver's cab and

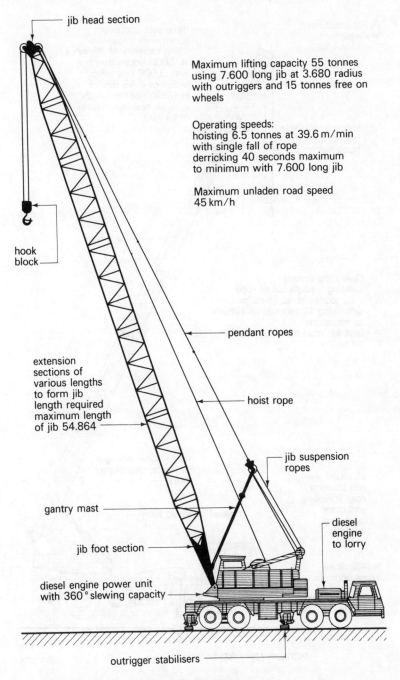

jib head section

Maximum lifting capacity 55 tonnes
using 7.600 long jib at 3.680 radius
with outriggers and 15 tonnes free on
wheels

Operating speeds:
hoisting 6.5 tonnes at 39.6 m/min
with single fall of rope
derricking 40 seconds maximum
to minimum with 7.600 long jib

Maximum unladen road speed
45 km/h

hook
block

pendant ropes

extension
sections of
various lengths
to form jib
length required
maximum length
of jib 54.864

hoist rope

jib suspension
ropes

gantry mast

diesel
engine
to lorry

jib foot section

diesel engine power unit
with 360° slewing capacity

outrigger stabilisers

Fig VII.23 Typical lorry mounted crane (Coles Cranes Ltd)

233

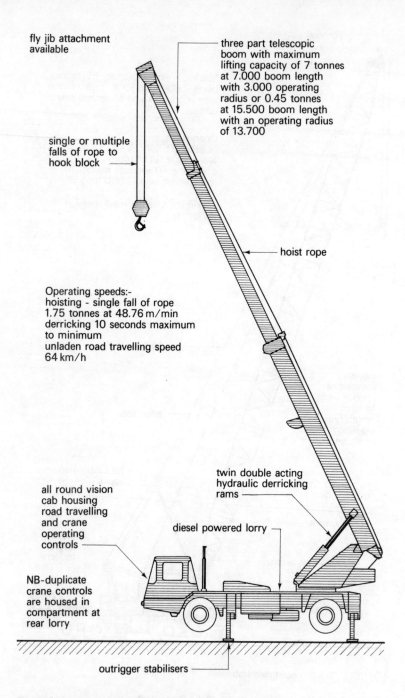

fly jib attachment
available

single or multiple
falls of rope to
hook block

three part telescopic
boom with maximum
lifting capacity of 7 tonnes
at 7.000 boom length
with 3.000 operating
radius or 0.45 tonnes
at 15.500 boom length
with an operating radius
of 13.700

hoist rope

Operating speeds:-
hoisting - single fall of rope
1.75 tonnes at 48.76 m/min
derricking 10 seconds maximum
to minimum
unladen road travelling speed
64 km/h

twin double acting
hydraulic derricking
rams

all round vision
cab housing
road travelling
and crane
operating
controls

diesel powered lorry

NB-duplicate
crane controls
are housed in
compartment at
rear lorry

outrigger stabilisers

Fig VII.24 Lorry mounted telescopic crane (Coles Cranes Ltd)

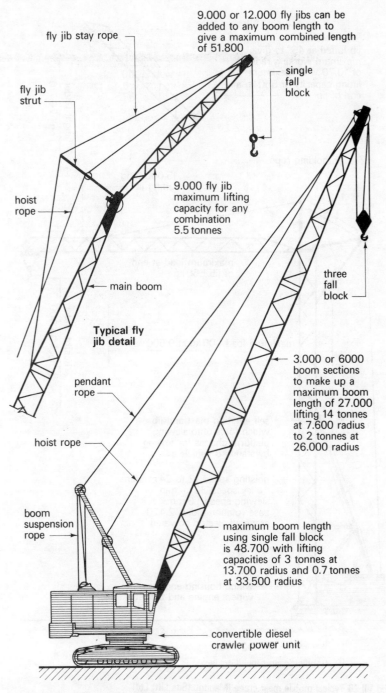

fly jib stay rope

9.000 or 12.000 fly jibs can be added to any boom length to give a maximum combined length of 51.800

single fall block

fly jib strut

hoist rope

9.000 fly jib maximum lifting capacity for any combination 5.5 tonnes

three fall block

main boom

Typical fly jib detail

pendant rope

3.000 or 6000 boom sections to make up a maximum boom length of 27.000 lifting 14 tonnes at 7.600 radius to 2 tonnes at 26.000 radius

hoist rope

boom suspension rope

maximum boom length using single fall block is 48.700 with lifting capacities of 3 tonnes at 13.700 radius and 0.7 tonnes at 33.500 radius

convertible diesel crawler power unit

Fig VII.25 Typical track mounted crane (Thomas Smith & Sons Ltd)

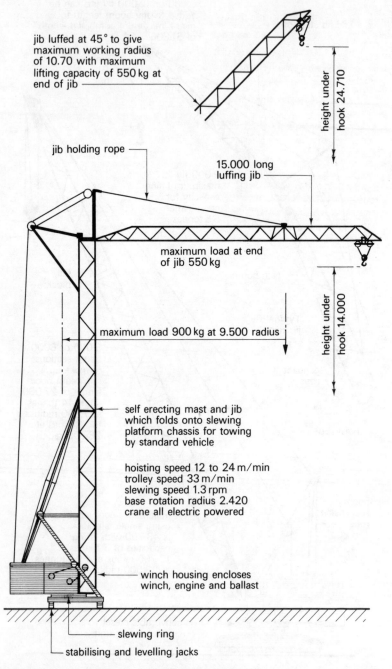

jib luffed at 45° to give maximum working radius of 10.70 with maximum lifting capacity of 550 kg at end of jib

height under hook 24.710

jib holding rope

15.000 long luffing jib

maximum load at end of jib 550 kg

height under hook 14.000

maximum load 900 kg at 9.500 radius

self erecting mast and jib which folds onto slewing platform chassis for towing by standard vehicle

hoisting speed 12 to 24 m/min
trolley speed 33 m/min
slewing speed 1.3 rpm
base rotation radius 2.420
crane all electric powered

winch housing encloses winch, engine and ballast

slewing ring

stabilising and levelling jacks

Fig VII.26 Typical mobile mast crane (Manitou (Site Lift) Ltd)

hook supporting saddle. The beam is supported by rail-mounted 'A' frames on powered bogies situated on both sides of the building under construction. This is a particularly safe form of crane as it requires no ballast, gives the driver an excellent all-round view and allows the hook three-way movement of vertical, horizontal and transverse directions. Although limited in application this special form of mobile crane can be very usefully and economically employed on repetitive and partially prefabricated blocks of medium-rise dwellings.

Static or stationary cranes

These cranes are fixed at their working position and are used primarily for lifting heavy loads such as structural steelwork. Although more common on civil engineering contracts they can be successfully and economically employed by building contractors.

Guyed derrick: simple and inexpensive form of static crane powered by a diesel engine or electric motor, consisting of a lattice mast with a pedestal bearing stabilised by five or more anchored guy ropes. The jib is of the low pivot type and is slightly shorter in length than the mast height so that it can rotate through the whole $360°$ without fouling the guy ropes if raised in the near vertical position — see Fig. VII.27 for typical details.

Scotch derrick: consists of a slewing mast and a luffing jib which is usually longer than the jib used on a similar capacity guyed derrick. Stabilisation of a Scotch derrick is obtained by using lattice members called guys and stays. Two guys are fixed to the top of the slewing mast at an angle of $45°$ with the horizontal and at an angle of $90°$ in plan, the lower ends of the guys are connected to the ends of horizontal stays fixed to base of the mast forming an angle of $90°$ in plan. A horizontal brace is fixed between the ends of the guys and stays, forming a complete triangulation of the stabilising members together with the mast. Scotch derricks are only capable of slewing $270°$, being restricted in further rotational movement by the sloping lattice guys. Resistance to overturning can be provided by kentledge applied to the struts and brace, or these members can be bolted to temporary concrete bases. The power can be supplied by a diesel engine but the common power source is an electric motor — for typical details see Fig. VII.27.

Monotower cranes: these are basically an elevated Scotch derrick crane consisting of a fabricated and well-braced tower surmounted by a derrick crane, the mast of which extends to a pivot bearing well down the height of the tower. The tower can be up to 60.000 in height with a job length of some 40.000 capable of raising 2 tonnes at its maximum radius. To be

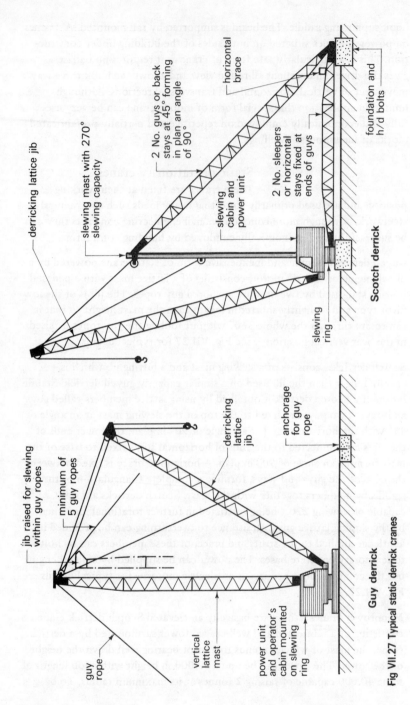

Fig VII.27 Typical static derrick cranes

The labels in the figure read:

Scotch derrick:
- horizontal brace
- 2 No. guys or back stays at 45° forming in plan an angle of 90°
- 2 No. sleepers or horizontal stays fixed at ends of guys
- foundation and h/d bolts
- slewing mast with 270° slewing capacity
- slewing cabin and power unit
- slewing ring
- derricking lattice jib

Guy derrick:
- jib raised for slewing within guy ropes
- minimum of 5 guy ropes
- derricking lattice jib
- anchorage for guy rope
- guy rope
- vertical lattice mast
- power unit and operator's cabin mounted on slewing ring

economic this form of crane needs to be centrally sited to give maximum site coverage.

Tower cranes

Since their introduction in 1950 by the then Department of Scientific and Industrial Research the tower crane has been universally accepted by the building industry as a standard piece of plant required for construction of medium- to high-rise structures. These cranes are available in several forms with a horizontal jib carrying a saddle or a trolley, or alternatively with a luffing or derricking jib with a lifting hook at its extreme end. Horizontal jibs can bring the load closer to the tower whereas luffing jibs can be raised to clear obstructions such as adjacent building, an advantage on confined sites. The basic types of tower cranes available are:

1. Self-supporting static tower cranes.
2. Supported static tower cranes.
3. Travelling tower cranes.
4. Climbing cranes.

Self-supporting static tower cranes: these cranes generally have a greater lifting capacity than other types of crane. The mast of the self-supporting tower crane must be firmly anchored at ground level to a concrete base with holding down bolts or alternatively to a special mast base section cast into a foundation. They are particularly suitable for confined sites and should be positioned in front or to one side of the proposed building with a jib of sufficient length to give overall coverage of the new structure. Generally these cranes have a static tower but types with a rotating or slewing tower and luffing jib are also available — see Fig. VII.28 for typical self-supporting crane example.

Supported static tower cranes: these are similar in construction to self-supporting tower cranes but are used for lifting to a height in excess of that possible with self-supporting or travelling tower cranes. The tower or mast is fixed or tied to the structure using single or double steel stays to provide the required stability. This tying back will induce stresses in the supporting structure which must therefore be of adequate strength. Supported tower cranes usually have horizontal jibs since the rotation of a luffing jib mast renders it as unsuitable for this application — see Fig. VII.29 for typical example.

Travelling tower cranes: to obtain better site coverage with a tower crane a rail-mounted or travelling crane could be used. The crane travels on heavy wheeled bogies mounted on a wide gauge (4.200) rail track with gradients

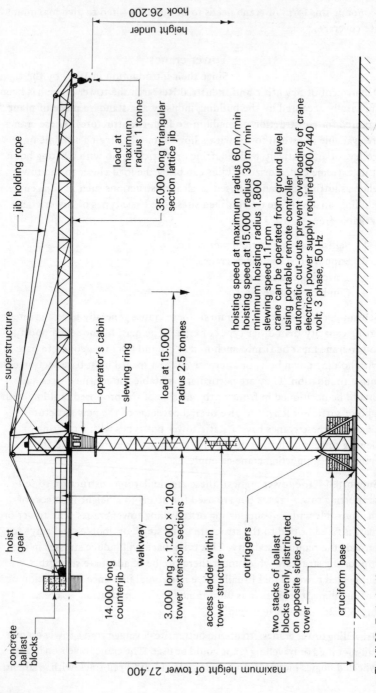

height under hook 26.200

load at maximum radius 1 tonne

35.000 long triangular section lattice jib

jib holding rope

superstructure

operator's cabin

slewing ring

load at 15.000 radius 2.5 tonnes

hoisting speed at maximum radius 60 m/min
hoisting speed at 15.000 radius 30 m/min
minimum hoisting radius 1.800
slewing speed 1.1 rpm
crane can be operated from ground level
using portable remote controller
automatic cut-outs prevent overloading of crane
electrical power supply required 400/440
volt, 3 phase, 50 Hz

hoist gear

concrete ballast blocks

14.000 long counterjib

walkway

3.000 long × 1.200 × 1.200 tower extension sections

access ladder within tower structure

outriggers

two stacks of ballast blocks evenly distributed on opposite sides of tower

cruciform base

maximum height of tower 27.400

Fig VII.28 Typical self supporting static tower crane (Stotherd & Pitt Ltd)

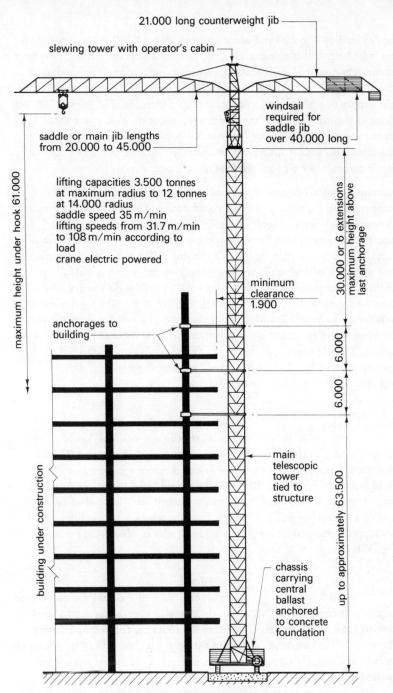

21.000 long counterweight jib

slewing tower with operator's cabin

windsail required for saddle jib over 40.000 long

saddle or main jib lengths from 20.000 to 45.000

lifting capacities 3.500 tonnes at maximum radius to 12 tonnes at 14.000 radius
saddle speed 35 m/min
lifting speeds from 31.7 m/min to 108 m/min according to load
crane electric powered

maximum height under hook 61.000

30.000 or 6 extensions maximum height above last anchorage

minimum clearance 1.900

anchorages to building

6.000

6.000

main telescopic tower tied to structure

building under construction

up to approximately 63.500

chassis carrying central ballast anchored to concrete foundation

Fig VII.29 Typical supported static tower crane (Babcock Weitz)

241

not exceeding 1 in 200 and curves not less than 11.000 radius depending on mast height. It is essential that the base for the railway track sleepers is accurately prepared, well drained, regularly inspected and maintained if the stability of the crane is to be ensured. The motive power is electricity, the supply of which should be attached to a spring loaded drum which will draw in the cable as the crane reverses to reduce the risk of the cable becoming cut or trapped by the wheeled bogies. Travelling cranes can be supplied with similar lifting capacities and jib arrangements as given for static cranes – see Fig. VII.30 for typical example.

Climbing cranes: design for tall buildings being located within and supported by the structure under construction. The mast which extends down through several storeys requires only a small (1.500 to 2.000 square) opening in each floor. Support is given at floor levels by special steel collars, frames and wedges. The raising of the static mast is carried out using a winch which is an integral part of the system. Generally this form of crane requires a smaller horizontal or luffing jib to cover the construction area than a static or similar tower crane. The jib is made from small, easy-to-handle sections which are lowered down the face of the building, when the crane is no longer required, by means of a special winch attached to one section of the crane. The winch is finally lowered to ground level by hand when the crane has been dismantled – see Fig. VII.31 for typical crane details.

Crane skips and slings

Cranes are required to lift all kinds of materials ranging from prefabricated components to loose and fluid materials. Various skips or containers have been designed to carry loose or fluid materials – see Fig. VII.32. Skips should be of sound construction, easy to attach to the crane hook, easily cleaned, easy to load and unload and of a suitable capacity. Prefabricated components are usually hoisted from predetermined lifting points by using wire or chain slings – see Fig. VII.32.

Wire ropes

Wire ropes consist of individual wires twisted together to form strands which are then twisted together around a steel core to form a rope or cable. Ordinary lay ropes are formed by twisting the wires in the individual strands in the opposite direction to the group of strands whereas in lang lay ropes the wires in the individual strands are twisted in the same direction as the groups of strands. Lang lay ropes generally have better wearing properties due to the larger surface area of

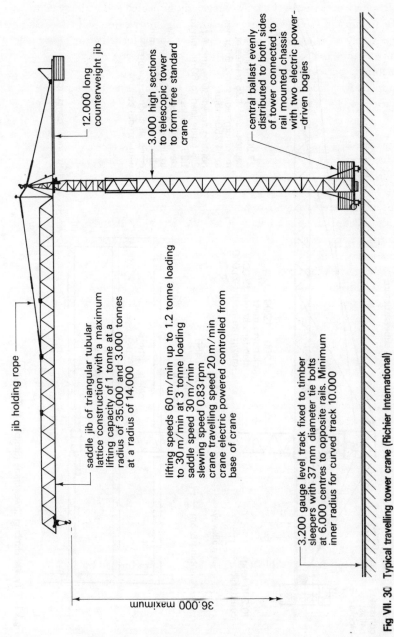

12.000 long counterweight jib

3.000 high sections to telescopic tower to form free standard crane

central ballast evenly distributed to both sides of tower connected to rail mounted chassis with two electric power-driven bogies

jib holding rope

saddle jib of triangular tubular lattice construction with a maximum lifting capacity of 1 tonne at a radius of 1.000 and 3.000 tonnes at a radius of 14.000

lifting speeds 60 m/min up to 1.2 tonne loading to 30 m/min at 3 tonne loading
saddle speed 30 m/min
slewing speed 0.83 rpm
crane travelling speed 20 m/min
crane electric powered controlled from base of crane

3.200 gauge level track fixed to timber sleepers with 37 mm diameter tie bolts at 6.000 centres to opposite rails. Minimum inner radius for curved track 10.000

36.000 maximum

Fig VII. 30 Typical travelling tower crane (Richier International)

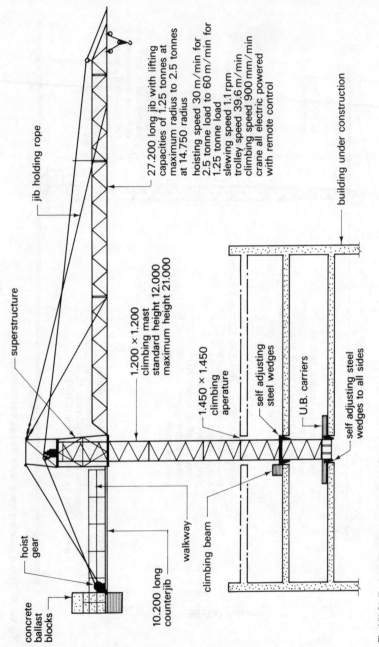

jib holding rope

27.200 long jib with lifting
capacities of 1.25 tonnes at
maximum radius to 2.5 tonnes
at 14.750 radius

hoisting speed 30 m/min for
2.5 tonne load to 60 m/min for
1.25 tonne load
slewing speed 1.1 rpm
trolley speed 39.6 m/min
climbing speed 900 mm/min
crane all electric powered
with remote control

building under construction

superstructure

1.200 × 1.200
climbing mast
standard height 12.000
maximum height 21.000

1.450 × 1.450
climbing aperature

self adjusting
steel wedges

U.B. carriers

self adjusting steel
wedges to all sides

walkway

climbing beam

hoist
gear

concrete
ballast
blocks

10.200 long
counterjib

Fig VII.31 Typical climbing tower crane (Stothert & Pitt Ltd)

244

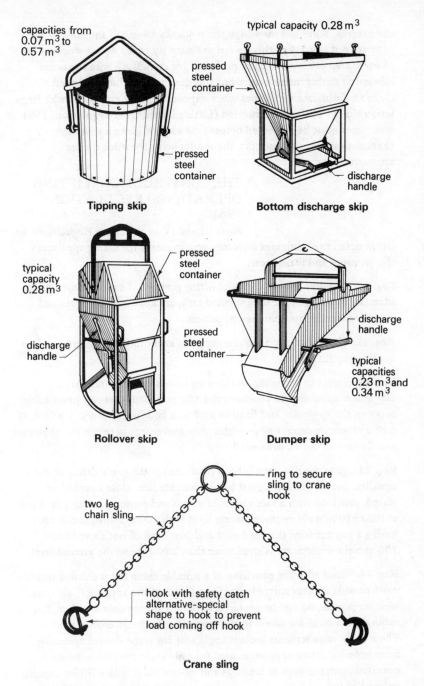

capacities from
0.07 m³ to
0.57 m³

pressed
steel
container

Tipping skip

typical capacity 0.28 m³

pressed
steel
container

discharge
handle

Bottom discharge skip

typical
capacity
0.28 m³

pressed
steel
container

discharge
handle

Rollover skip

pressed
steel
container

discharge
handle

typical
capacities
0.23 m³ and
0.34 m³

Dumper skip

ring to secure
sling to crane
hook

two leg
chain sling

hook with safety catch
alternative-special
shape to hook to prevent
load coming off hook

Crane sling

Fig VII.32 Typical crane skips and sling

the external wires but they have the tendency to spin if the ends of the rope are not fixed. For this reason ordinary lay ropes with a working life of up to two years are usually preferred for cranes. All wire ropes are lubricated during manufacture but this does not preclude the need to clean and lubricate wire ropes when exposed to the elements. Under Regulation No. 34 of the Construction (Lifting Operations) Regulations 1961 wire ropes must be inspected before first use and given a thorough examination every six months, the results being recorded on the appropriate form.

THE CONSTRUCTION (LIFTING OPERATIONS) REGULATIONS 1961

Parts III and IV of the above Regulations set out in detail the minimum statutory requirements for lifting appliances, chains, ropes and lifting gear.

Reg. 10: requires that all forms of lifting gear are of sound construction, adequate in strength for the intended task, are kept in good order and inspected weekly by a competent person.

Reg. 11: deals with the adequate support, anchoring, fixing and erection of lifting appliances.

Reg. 12: refers to travelling and slewing cranes and requires that a 600 mm wide minimum clearance must be provided wherever practicable between the appliance and fixtures such as a building or access scaffold. If such a clearance cannot be provided then movement between the appliance and fixture should be prevented.

Reg. 13: gives specific details for a platform for the crane driver or the signaller. Such a platform must be of adequate size, close boarded or plated, provided with a safe means of access and protected with guard rails at least 910 mm above the platform level with 200 mm high toe boards having a gap between the toe board and guard rail of not more than 700 mm if the platform is sited more than 2.000 above the ground level.

Reg. 14: deals with the provision of a suitable cabin for the driver which must provide an unrestricted view for safe use of the appliance, adequate weather protection and be heated when in use during cold weather. The cabin must also allow access to the machinery for maintenance work. These cabin requirements are not applicable for crane drivers operating from indoors, lifting appliances used for only short durations, hoists operated from the cage or landings and mobile plant with a lifting capacity below 1 tonne.

Reg. 15: gives the requirements for the suitability of drums and pulleys.

Reg. 16: sets out the provisions for brakes, controls and safety devices, the need for clear marking of controls and designed to prevent accidental operation.

Reg. 17: deals with safe means of access for the purposes of examination, repair or lubrication, particularly where a person can fall more than 2.000.

Reg. 18: gives strength and fixing requirements for poles and beams supporting pulley blocks or gin wheels.

Reg. 19: deals with the stability of lifting appliances when used on soft ground, uneven surfaces and slopes. The crane must be either anchored to the ground or to a foundation, or suitably counterweighted to prevent overturning.

Reg. 20: is concerned with rail-mounted cranes and the need for the track to be laid and secured on a firm foundation to prevent the risk of derailment. This Regulation also deals with the requirements for buffers, effective braking systems and adequate maintenance of both track and equipment.

Reg. 21: sets out the strength requirements for the mounting of cranes on bogies, trolley or wheeled carriages.

Reg. 22: deals with cranes having a derricking jib operated through a clutch which must have an effective interlock arrangement between the derricking clutch and the pawl and the ratchet on the derricking drum.

Reg. 23: restricts the use of cranes to direct raising and lowering operations. Cranes with a derricking jib shall not be used with the jib at a radius greater than that specified on the test certificate.

Reg. 24: no crane which has any timber structural member shall be used.

Reg. 25: cranes must always be erected under the supervision of a competent person.

Regs. 26 and 27: covers the requirements of persons operating lifting appliances and signalling requirements. Drivers of lifting appliances must be trained, experienced and over 18 years of age. If the driver does not have a clear vision during the whole of the lifting operation an adequate signalling system must be used. A signaller must be over 18 years of age, capable of giving clear and distinct signals by hand, mechanical or electrical means.

Reg. 28: sets out the testing, examination and inspection requirements. Cranes, grabs and winches – testing and thorough examination every four years, thorough examination every 14 months and inspections to be carried out weekly. Pulley blocks, gin wheels and sheer legs — testing and thorough examination before first use, thorough examination every 14 months and weekly inspections.

Reg. 29: requires that all cranes are clearly marked with their safe maximum working loads relevant to lifting radius and maximum operating radius when fitted with a derricking jib.

Reg. 30: jib cranes must be fitted with an automatic safe load indicator approved by the Chief Inspector of Factories such as a warning light for the driver and a warning bell for persons nearby.

Reg. 31: except for testing purposes the safe working load must not be exceeded.

Reg. 32: when loads being lifted are approaching the safe maximum load the initial lift should be short. A check should then be made to establish safety and stability before proceeding to complete the lift.

Reg. 33: gives details of the stability requirements for Scotch and guy derricks.

Regs. 34 to 41: gives the special requirements for lifting gear such as chains, slings, ropes, hooks, chain shackles and eye bolts. Testing and examination requirements of before each use and thorough examination every six months are also set out in these Regulations.

Apart from the legal requirements summarised above commonsense precautions on site must be taken such as the clear marking of high voltage electric cables and leaving the crane in an 'out of service' position when unattended or if storm or high wind conditions prevail. Most tower cranes can operate in wind conditions of up to 60 km/h. The usual 'out of service' position for tower cranes is as follows:

1. Jibs to be left on free slew and pointed in the direction of the wind on the leeward side of the tower.
2. Fuel and power supplies switched off.
3. Load removed.
4. Hook raised to highest position.
5. Hook positioned close to tower.
6. Rail-mounted cranes should have their wheels chocked or clamped.

ERECTION OF CRANES

Before commencing to erect a crane careful consideration must be given to its actual position on the site. Like all forms of plant maximum utilisation is the ultimate aim, therefore a central

248

position within reach of all storage areas, loading areas and activity areas is required. Generally output will be in the region of 18 to 20 lifts per hour therefore the working sequence of the crane needs to be carefully planned and co-ordinated if full advantage is to be made of the crane's capabilities.

The erection of mast and tower cranes varies with the different makes but there are several basic methods. Mast cranes are usually transported in a collapsed and folded position and are quickly unfolded and erected on site, using built-in lifting and erection gear. Tower cranes, however, have to be assembled on site. In some cases the superstructure which carries the jib and counterjib is erected on the base frame. The top section of the tower or pintle is raised by internal climbing gear housed within the superstructure; further 3.000 tower lengths can be added as the pintle is raised until the desired tower height has been reached. The jib and counterjib are attached at ground level to the superstructure which is then raised to the top of the pintle; this whole arrangement then slews around the static tower.

Another method of assembly and erection adopted by some manufacturers is to raise the first tower section onto a base, assemble the jib and counterjib and fix these to the first tower section. Using the facilities of the jib, further tower sections can be fitted inside the first section and elevated on a telescopic principle, this procedure being repeated until the desired height has been reached. A similar approach to the last method is to have the jib and top tower section fixed to a cantilever bracket arrangement so that it is offset from the main tower. Further sections can be added to the assembly until the required height is reached when the jib assembly can be transferred to the top of the tower.

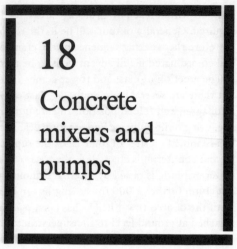

18
Concrete mixers and pumps

The mixing and transporting of concrete and mortar mixes are important activities on most building sites from the very small to the very large contract. The choice of method for mixing and transporting the concrete or mortar must be made on the basis of the volume of mixed material required in any given time and also the horizontal and vertical transportation distances involved. Consideration must also be given to the use of ready mixed concrete especially where large quantities are required and/or site space is limited.

CONCRETE MIXERS

Most concrete mixers used on building sites are of the batch type conforming to the minimum recommendations of BS 1305 which defines two basic forms, namely the drum type or free-fall concrete mixer and the pan type or forced action concrete mixer. The drum mixers are subdivided into three distinct forms:

1. *Tilting drum* (T) — in which the single-compartment drum has an inclinable axis with loading and discharge through the front opening. This form of mixer is primarily intended for small batch outputs ranging from 100T to 200T litres. It should be noted that mixer output capacities are given in litres for sizes up to 1 000 litres and in cubic metres for outputs over 1 000 litres, a letter suffix designating the type being also included in the title. In common with all drum mixers tilting mixers have fixed blades inside the revolving drum which lift the mixture and at a certain point in each revolution allow

the mixture to drop towards the bottom of the drum to recommence the mixing cycle. The complete cycle time for mixing one batch from load to reload is usually specified as 2½ minutes. Typical examples of tilting drum mixers are shown in Fig. VII.33.

2. *Non-tilting drum* (NT) — in which the single-compartment drum has two openings and rotates on a horizontal axis with output capacities ranging from 200 NT to 750 NT. Loading is through the front opening and discharge through the rear opening by means of a discharge chute collecting the mixture from the top of the drum. The chute should form an angle of not less than 40° with the horizontal axis of the drum.

3. *Reversing drum* (R) — a more popular version of a mixer with a drum rotating on a horizontal axis than the non-tilting drum mixer described above. Capacities of this type of mixer range from 200R to 500R. Loading is through a front opening and discharge from a rear opening carried out by reversing the rotation of the drum — see Fig. VII.34.

Generally mixers with an output capacity exceeding 200 litres are fitted with an automatic or manually operated water system which will deliver a measured volume of water to the drum of the mixer. Table 1 of BS 1305 gives recommended minimum water tank capacities for the various mixer sizes.

Forced action mixers (P) are generally for larger capacity outputs than the drum mixers described above and can be obtained within the range of 200 P to P2.0. The mixing of the concrete is achieved by the relative movements between the mix, pan and blades or paddles. Usually the pan is stationary whilst the paddles or blades rotate but rotating pan models consisting of a revolving pan and a revolving mixer blade or star giving a shorter mixing time of 30 seconds with large outputs are also available. In general pan mixers are not easily transported and for this reason are usually only employed on large sites where it would be an economic proposition to install this form of mixer.

CEMENT STORAGE

Cement for the mixing of mortars or concrete can be supplied in 50 kg bags or in bulk for storage on site prior to use. Bagged cement requires a dry and damp-free store to prevent air setting taking place (see Chapter 1, Volume 2). If large quantities of cement are required an alternative method of storage is the silo which will hold cement supplied in bulk under ideal conditions. A typical cement silo consists of an elevated welded steel cylindrical container supported on

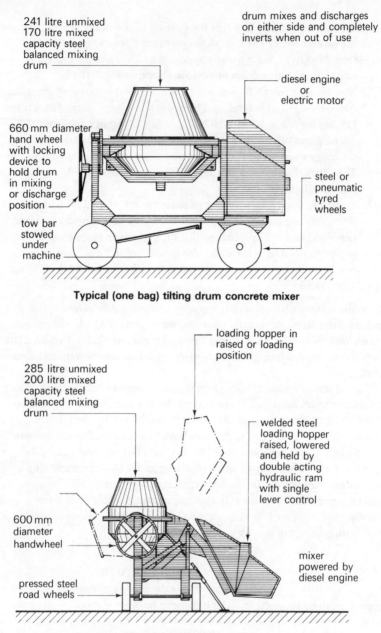

241 litre unmixed
170 litre mixed
capacity steel
balanced mixing
drum

drum mixes and discharges
on either side and completely
inverts when out of use

diesel engine
or
electric motor

660 mm diameter
hand wheel
with locking
device to
hold drum
in mixing
or discharge
position

steel or
pneumatic
tyred
wheels

tow bar
stowed
under
machine

Typical (one bag) tilting drum concrete mixer

loading hopper in
raised or loading
position

285 litre unmixed
200 litre mixed
capacity steel
balanced mixing
drum

welded steel
loading hopper
raised, lowered
and held by
double acting
hydraulic ram
with single
lever control

600 mm
diameter
handwheel

mixer
powered by
diesel engine

pressed steel
road wheels

Typical hopper fed tilting concrete mixer

Fig VII.33 Typical tilting drum mixers (Liner Concrete Machinery Co Ltd)

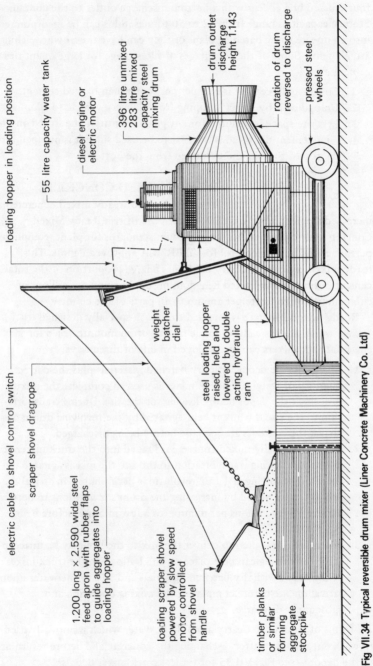

electric cable to shovel control switch

scraper shovel dragrope

1.200 long × 2.590 wide steel feed apron with rubber flaps to guide aggregates into loading hopper

loading scraper shovel powered by slow speed motor controlled from shovel handle

timber planks or similar forming aggregate stockpile

55 litre capacity water tank

loading hopper in loading position

diesel engine or electric motor

weight batcher dial

steel loading hopper raised, held and lowered by double acting hydraulic ram

396 litre unmixed 283 litre mixed capacity steel mixing drum

drum outlet discharge height 1.143

rotation of drum reversed to discharge

pressed steel wheels

Fig VII.34 Typical reversible drum mixer (Liner Concrete Machinery Co. Ltd)

four crossed braced legs with a bottom discharge outlet to the container. Storage capacities range from 12 to 50 tonnes. Silos can be incorporated into an on-site static batching plant or they can have their own weighing attachments. Some of the advantages of silo storage for large quantities of cement are:

1. Cost of bulk cement is cheaper per tonne than bagged cement.
2. Unloading is by direct pumping from delivery vehicle to silo.
3. Less site space is required for any given quantity to be stored on site.
4. First cement delivered is the first to be used since it is pumped into the top of the silo and extracted from the bottom.

READY MIXED CONCRETE

The popularity of ready mixed concrete has increased tremendously since 1968 when the British Ready Mixed Concrete Association laid down minimum standards for plant, equipment, personnel and quality control for all BRMCA approved depots. The ready mixed concrete industry consumes a large proportion of the total cement output of the United Kingdom in supplying many millions of cubic metres of concrete per annum to all parts of the country.

Ready mixed concrete is supplied to sites in specially designed truck mixers which are basically a mobile mixing drum mounted on a lorry chassis. Truck mixers can be employed in one of three ways:

1. Loaded at the depot with dry batched materials plus the correct quantity of water, the truck mixer is used to complete the mixing process at the depot before leaving for the site. During transportation to the site the mix is kept agitated by the revolving drum; on arrival the contents are remixed before being discharged.
2. Fully or partially mixed concrete is loaded into the truck mixer at the depot. During transportation to the site the mix is agitated by the drum revolving at 1 to 2 revolutions per minute. On arrival the mix is finally mixed by increasing the drum's revolutions to between 10 and 15 revolutions per minute for a few minutes before being discharged.
3. When the time taken to deliver the mix to the site may be unacceptable the mixing can take place on site by loading the truck mixer at the depot with dry batched materials and adding the water upon arrival on site before completing the mixing operation and subsequent discharge.

All forms of truck mixer carry a supply of water which is normally used to wash out the drum after discharging the concrete and before returning to the depot — see Fig. VII.35 for typical truck mixer details.

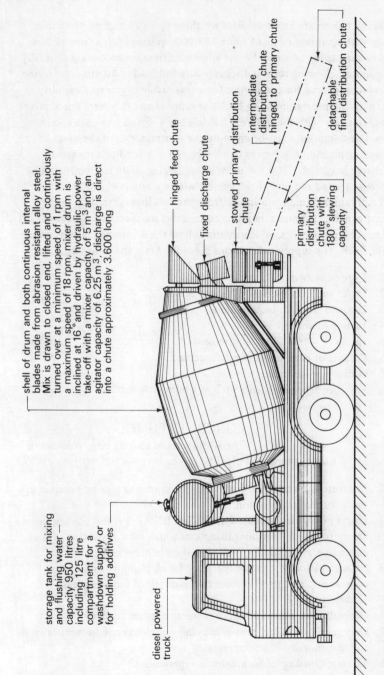

shell of drum and both continuous internal
blades made from abrasion resistant alloy steel.
Mix is drawn to closed end, lifted and continuously
turned over at a minimum speed of 1 rpm with
a maximum speed of 18 rpm, mixer drum is
inclined at 16° and driven by hydraulic power
take-off with a mixer capacity of 5 m³ and an
agitator capacity of 6.25 m³, discharge is direct
into a chute approximately 3.600 long

hinged feed chute

fixed discharge chute

stowed primary distribution
chute

intermediate
distribution chute
hinged to primary chute

detachable
final distribution chute

primary
distribution
chute with
180° slewing
capacity

storage tank for mixing
and flushing water —
capacity 950 litres
including 125 litre
compartment for a
washdown supply or
for holding additives

diesel powered
truck

Fig VII.35 Typical ready mix concrete truck details

Truck mixers are heavy vehicles weighing up to 24 tonnes when fully laden with a turning circle of some 15.000 requiring both a firm surface and turning space on site. The site allowance time for unloading is usually 30 minutes, allowing for the discharge of a full load in 10 minutes leaving 20 minutes of free time to permit for a reasonable degree of flexibility in planning and programming to both the supplier and the user. Truck mixer capacities vary with the different models but 4, 5 and 6 m^3 are common sizes. Consideration must be given by the contractor as to the best unloading position since most truck mixers are limited to a maximum discharge height of 1.500 and using a discharging chute to a semi-circular coverage around the rear of the vehicle within a radius of 3.000.

To obtain maximum advantage from the facilities offered by ready mixed concrete suppliers, building contractors must place a clear order of the exact requirements, which should follow the recommendations given in BS 5328. The supply instructions should contain the following:

1. Type of cement.
2. Types and maximum sizes of aggregates.
3. Test and strength requirements.
4. Testing methods.
5. Slump or workability requirements.
6. Volume of each separate mix specified.
7. Delivery programme.
8. Any special requirements such as a pumpable mix.

CONCRETE PUMPS

The advantages of moving large volumes of concrete by using a pump and pipeline can be listed as follows:

1. Concrete is transported from point of supply to placing position in one continuous operation.
2. Faster pours can be achieved with less labour. Typical placing figures are up to 100 m^3 per hour using a two-man crew consisting of the pump operator and an operator at the discharge end.
3. No segregation of mix is experienced with pumping and a more consistent placing and compaction is obtained requiring less vibration.
4. Generally site plant and space requirements are reduced.
5. Only method available for conveying wet concrete both vertically and horizontally in one operation.
6. No shock loading of formwork is experienced.
7. Generally the net cost of placing concrete is reduced.

Against the above listed advantages must be set the following limitations:

1. Concrete supply must be consistent and regular, which can usually be achieved by well-planned and organised deliveries of ready mixed concrete. It should be noted that under ideal conditions the discharge rate of each truck mixer can be in the order of 10 minutes.
2. Concrete mix must be properly designed and controlled since not all concrete mixes are pumpable. The concrete is pumped under high pressure which can cause bleeding and segregation of the mix; therefore the mix must be properly designed to avoid these problems as well as having good cohesive, plasticity and self-lubricating properties to enable it to be pumped through the system without excessive pressure and without causing blockages.
3. More formwork will be required to receive the high output of the pump to make its use an economic proposition.

Most pumps used today are of the twin-cylinder hydraulically driven design either as a trailer pump or lorry-mounted pump using a small bore (100 mm diameter) pipeline capable of pumping concrete 85.000 vertically and 200.000 horizontally although these figures will vary with the actual pump used — see Fig. VII.36 for typical example. The delivery pipes are usually of rigid seamless steel in 3.000 lengths except where flexibility is required as on booms and at the delivery end. Large radius bends of up to 1.000 radius giving 22½°, 45° and 90° turning are available to give flexible layout patterns. Generally small diameter pipes of 75 and 100 mm are used for vertical pumping whereas larger diameters of up to 150 mm are used for horizontal pumping. If a concrete mix with large aggregates is to be pumped the pipe diameter should be at least three or four times the maximum aggregate size.

The time required on site to set up a pump is approximately 30 to 45 minutes. The pump operator will require a supply of water and grout for the initial coating of the pipeline, this usually requires about two or three bags of cement. A hard standing should be provided for the pump with adequate access and turning space for the attendant ready mixed concrete vehicles. The output of a concrete pump will be affected by the distance the concrete is to be pumped, therefore the pump should be positioned so that it is as close to the discharge point as is practicable. Pours should be planned so that they progress backwards towards the pump, removing the redundant pipe lengths as the work proceeds.

Generally if the volume of concrete to be placed is sufficient to warrant hiring a pump and operators it will result in an easier, quicker and usually cheaper operation than placing the concrete by the traditional method of

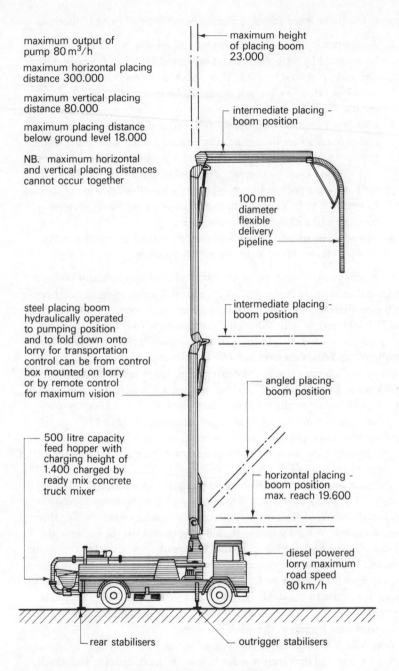

maximum output of
pump 80 m³/h

maximum horizontal placing
distance 300.000

maximum vertical placing
distance 80.000

maximum placing distance
below ground level 18.000

NB. maximum horizontal
and vertical placing distances
cannot occur together

maximum height
of placing boom
23.000

intermediate placing -
boom position

100 mm
diameter
flexible
delivery
pipeline

steel placing boom
hydraulically operated
to pumping position
and to fold down onto
lorry for transportation
control can be from control
box mounted on lorry
or by remote control
for maximum vision

intermediate placing -
boom position

angled placing-
boom position

500 litre capacity
feed hopper with
charging height of
1.400 charged by
ready mix concrete
truck mixer

horizontal placing -
boom position
max. reach 19.600

diesel powered
lorry maximum
road speed
80 km/h

rear stabilisers

outrigger stabilisers

Fig VII.36 Typical lorry mounted concrete pump (Schwing)

258

crane and skip with typical outputs of 15 to 20 m^3 per hour as opposed to the 60 to 100 m^3 per hour output of the concrete pump. Concrete pumping and placing demands a certain amount of skill and experience and for this reason most pumps in use are hired out and operated by specialist contractors.

19
Scaffolding

A scaffold is a temporary frame usually constructed from steel or aluminium alloy tubes clipped or coupled together to provide a means of access to high level working areas as well as providing a safe platform from which to work. The two basic forms of scaffolding, namely the putlog scaffold with its single row of uprights or standards set outside the perimeter of the building and partly supported by the structure and independent scaffolds which have two rows of standards should have been covered in the second year of study (see Chapter 4, Volume 2).

It is therefore only necessary to consider in this text the special scaffolds such as slung, suspended, truss-out and gantry scaffolds as well as the easy-to-erect system scaffolds. It cannot be over-emphasised that all scaffolds must fully comply with the minimum requirements set out in the Construction (Working Places) Regulations 1966 to which reference must be made (see Chapter 4, Volume 2).

Slung scaffolds: these are scaffolds which are suspended by means of wire ropes or chains and are not provided with a means of being raised or lowered by a lifting appliance. Their main use is for gaining access to high ceilings or the underside of high roofs. A secure anchorage must be provided for the suspension ropes and this can usually be achieved by using the structural members of the roof over the proposed working area. Any member selected to provide the anchorage point must be inspected to assess its adequacy. At least six evenly spaced suspension wire ropes or chains should be used and these must be adequately secured at both ends.

The working platform is constructed in a similar manner to conventional scaffolds, consisting of ledgers, transoms, and timber scaffold boards with the necessary guard rails and toe boards. Working platforms in excess of 2.400 x 2.400 plan size should be checked to ensure that the supporting tubular components are not being overstressed.

Truss-out scaffolds: these are a form of an independent tied scaffold which rely entirely on the building for support and are used where it is impossible or undesirable to erect a conventional scaffold from ground level. The supporting scaffolding structure which projects from the face of the building is known as the truss-out. Anchorage is provided by adjustable struts fixed internally between the floor and ceiling from which projects the cantilever tubes. Except for securing rakers only right-angle couplers should be used. The general format for the remainder of the scaffold is as used for conventional independent scaffolds — see Fig. VII.37.

Suspended scaffolds: these consist of a working platform suspended from supports such as outriggers which cantilever over the upper edge of a building and in this form are a temporary means of access to the face of a building for the purposes of cleaning and light maintenance work. Many new tall structures have suspension tracks incorporated in the fascia or upper edge beam, or a cradle suspension track is fixed to the upper surface of the flat roof on which is supported a manual or power trolley with retractable davit arms for supporting the suspended working platform or cradle. All forms of suspended cradles must conform with the minimum requirements set out in the Construction (Working Places) Regulations 1966 with regard to platform boards, guard rails and toe boards. Cradles may be single units or grouped together to form a continuous working platform; if grouped together they are connected to one another at their abutment ends with hinges which form a gap not exceeding 25 mm wide. Figure VII.38 shows typical suspended scaffold details.

Mobile tower scaffolds: these are used mainly by painters and maintenance staff to gain access to ceilings where it is advantageous to have a working platform which can be readily moved to a new position. The scaffold is basically a square tower constructed from scaffold tubes mounted on wheels fitted with brakes. Access is gained by means of a vertical ladder securely fixed to one side of the tower. A working platform complying with all the relevant regulations with a plan size of not less than 1.200 x 1.200 should be provided. To ensure complete stability the height of the tower should not exceed three-and-a-half times its least lateral dimension for internal work and three times its least lateral dimension for external work with a maximum height of 10.000 unless tied to the structure — see Fig. VII.39 for typical details.

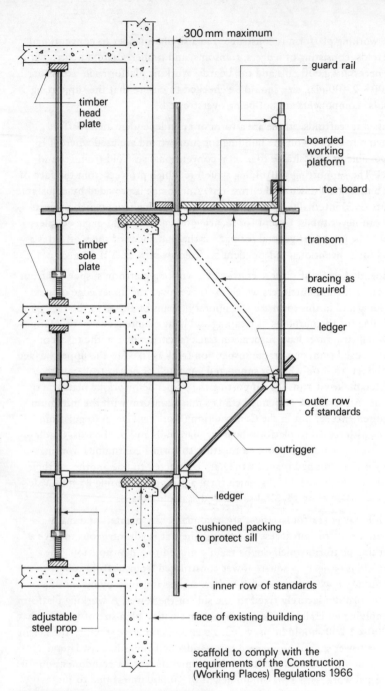

300 mm maximum

guard rail

timber head plate

boarded working platform

toe board

timber sole plate

transom

bracing as required

ledger

outer row of standards

outrigger

ledger

cushioned packing to protect sill

inner row of standards

adjustable steel prop

face of existing building

scaffold to comply with the requirements of the Construction (Working Places) Regulations 1966

Fig VII.37 Typical truss-out scaffold details

262

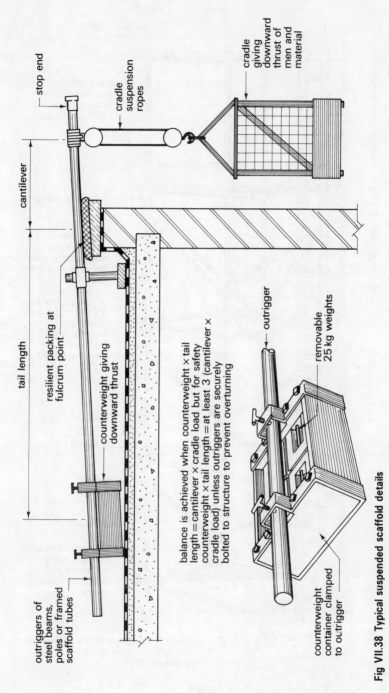

stop end

cradle suspension ropes

cradle giving downward thrust of men and material

cantilever

resilient packing at fulcrum point

tail length

counterweight giving downward thrust

outriggers of steel beams, poles or framed scaffold tubes

balance is achieved when counterweight × tail length = cantilever × cradle load but for safety counterweight × tail length = at least 3 (cantilever × cradle load) unless outriggers are securely bolted to structure to prevent overturning

outrigger

removable 25 kg weights

counterweight container clamped to outrigger

Fig VII.38 Typical suspended scaffold details

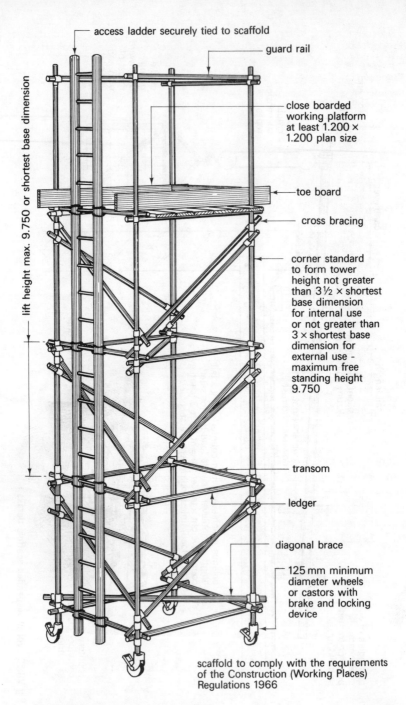

access ladder securely tied to scaffold

guard rail

close boarded
working platform
at least 1.200 ×
1.200 plan size

toe board

cross bracing

corner standard
to form tower
height not greater
than 3½ × shortest
base dimension
for internal use
or not greater than
3 × shortest base
dimension for
external use -
maximum free
standing height
9.750

transom

ledger

diagonal brace

125 mm minimum
diameter wheels
or castors with
brake and locking
device

lift height max. 9.750 or shortest base dimension

scaffold to comply with the requirements
of the Construction (Working Places)
Regulations 1966

Fig VII.39 Typical mobile tower scaffold

264

Birdcage scaffolds: these are used to provide a complete working platform at high level over a large area and consist basically of a two directional arrangement of standards, ledgers and transoms to support a close boarded working platform at the required height. To ensure adequate stability standards should be placed at not more than 2.400 centres in both directions and the whole arrangement adequately braced.

Gantries: these are forms of scaffolding used primarily as elevated loading and unloading platforms over a public footpath where the structure under construction or repair is immediately adjacent to the footpath. As in the case of hoardings local authority permission is necessary and their specific requirements such as pedestrian gangways, lighting and dimensional restrictions must be fully met. It may also be necessary to comply with police requirements as to when loading and unloading can take place. The gantry platform can also serve as a storage and accommodation area as well as providing the staging from which a conventional independent scaffold to provide access to the face of the building can be erected. Gantry scaffolds can be constructed from standard structural steel components as shown in Fig. VII.40 or from a system scaffold as shown in Fig. VII.41.

System scaffolds: these scaffolds are based upon the traditional independent steel tube scaffold but instead of being connected together with a series of loose couplers and clips they usually have integral interlocking connections. They are easy to erect, adaptable and generally they can be assembled and dismantled by semi-skilled labour. The design of these systems is such that the correct position of handrails, lift heights and all other aspects of the Construction (Working Places) Regulations 1966 are automatically met. Another advantage found in most of these system scaffolds is the elimination of internal cross-bracing, giving a clear walk through space at all levels; façade-bracing however may still be required. Fig. VII.42 shows details of a typical system scaffold and like the illustrations chosen for items of builders' mechanical plant is only intended to be representative of the many system scaffolds available.

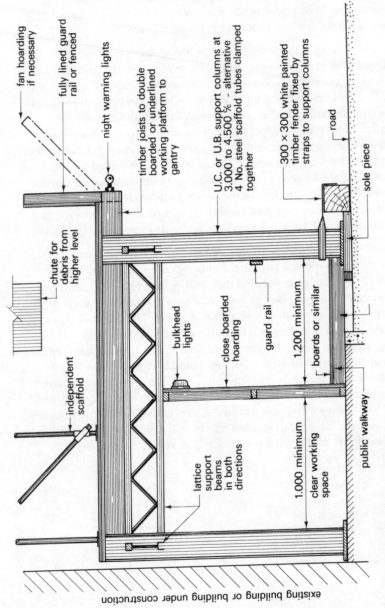

fan hoarding if necessary

fully lined guard rail or fenced

night warning lights

timber joists to double boarded or underlined working platform to gantry

U.C. or U.B. support columns at 3.000 to 4.500 % - alternative 4 No. steel scaffold tubes clamped together

300 × 300 white painted timber fender fixed by straps to support columns

road

sole piece

chute for debris from higher level

bulkhead lights

close boarded hoarding

guard rail

1.200 minimum

boards or similar

independent scaffold

public walkway

lattice support beams in both directions

1.000 minimum clear working space

existing building or building under construction

Fig VII.40 Typical gantry scaffold details 1

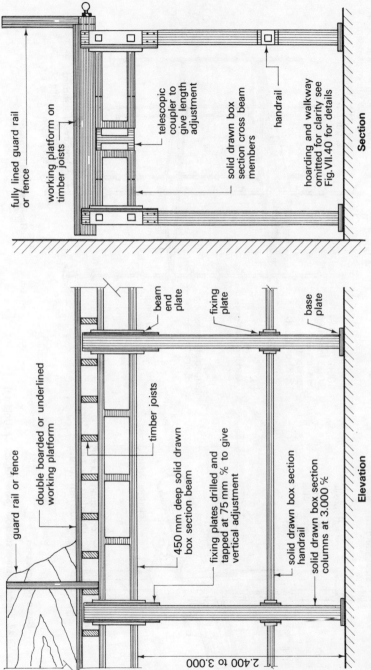

Section

- fully lined guard rail or fence
- working platform on timber joists
- telescopic coupler to give length adjustment
- solid drawn box section cross beam members
- handrail
- hoarding and walkway omitted for clarity see Fig.VII.40 for details

Elevation

- guard rail or fence
- double boarded or underlined working platform
- timber joists
- beam end plate
- fixing plate
- base plate
- 450 mm deep solid drawn box section beam
- fixing plates drilled and tapped at 75 mm % to give vertical adjustment
- solid drawn box section handrail
- solid drawn box section columns at 3.000 %
- 2.400 to 3.000

Fig VII.41 Typical gantry scaffold details 2

267

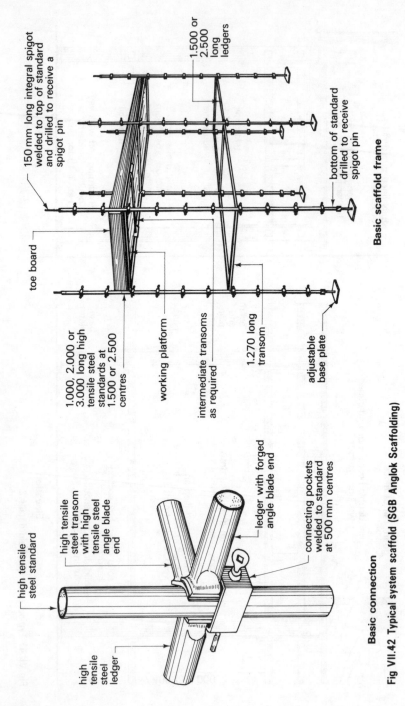

150 mm long integral spigot welded to top of standard and drilled to receive a spigot pin

1.500 or 2.500 long ledgers

bottom of standard drilled to receive spigot pin

Basic scaffold frame

toe board

1.000, 2.000 or 3.000 long high tensile steel standards at 1.500 or 2.500 centres

working platform

intermediate transoms as required

1.270 long transom

adjustable base plate

high tensile steel standard

high tensile steel transom with high tensile steel angle blade end

ledger with forged angle blade end

connecting pockets welded to standard at 500 mm centres

high tensile steel ledger

Basic connection

Fig VII.42 Typical system scaffold (SGB Anglok Scaffolding)

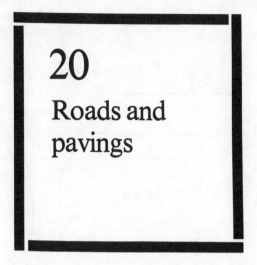

Part VIII
External works and internal slabs

20
Roads and pavings

A building contractor is not normally engaged to construct major roads or motorways as this is the province of the civil engineer. However, he can be involved in the laying of small estate roads, service roads and driveways and it is within this context that the student of construction technology would consider road works.

Before considering road construction techniques and types it is worth while considering some of the problems encountered by the designer when planning road layouts. The width of a road can be determined by the anticipated traffic flow which will use the road upon completion. The major layout problems occur at road junctions and at the termination of cul-de-sac roads. At right-angle junctions it is important that vehicles approaching the junction from any direction have a clear view of approaching vehicles intending to join the main traffic flow. Most planning authorities have layout restrictions at such junctions in the form of triangulated sight lines which give a distance and area in which an observer can see an object when both are at specific heights above the carriageway. Within this triangulated area street furniture or any other obstruction is not allowed — see Fig. VIII.1. Angled junctions by virtue of their distinctive layout do not present the same problems. At any junction a suitable radius should be planned so that vehicles filtering into the main road should not have to apply a full turning lock. The actual radius required will be governed by the anticipated vehicle types which will use the road.

Terminations at the end of cul-de-sacs must be planned to allow vehicles to turn round. For service roads this is usually based on the length and turning circle specifications of refuse collection vehicles which are

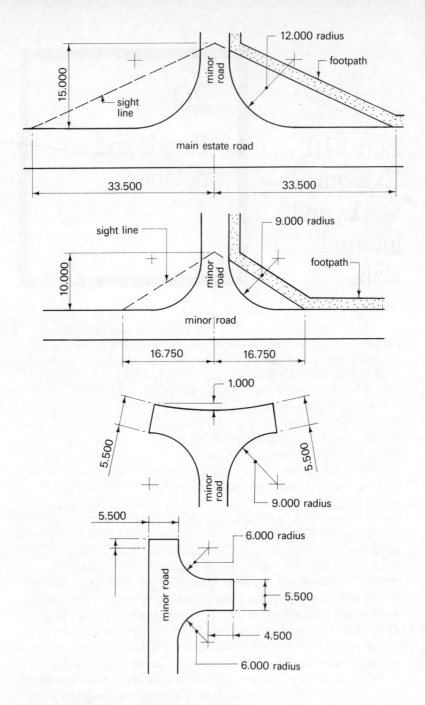

Fig VIII.1 Typical road junctions and terminations

probably the largest vehicles to use the road — typical examples are shown in Fig. VIII.1.

The construction of roads can be considered under two headings:

1. Preparation or earthworks.
2. Pavement construction.

Before any roadwork is undertaken a thorough soil investigation should be carried out to determine the nature of the subgrade which is the soil immediately below the topsoil that will ultimately carry the traffic loads from the pavement above. Soil investigations should preferably be carried out during the winter period when subgrade conditions will be at their worst. Trial holes should be taken down to at least 1.000 below the proposed formation level. The information required from these investigations to ensure that a good pavement design can be formulated are the elasticity, plasticity, cohesion and internal friction properties of the subgrade.

EARTHWORKS

This will include removing topsoil, scraping and grading the exposed surface to the required formation level, preparation of the subgrade to receive the pavement and forming any embankments and/or cuttings. The strength of the subgrade will generally decrease as the moisture content increases. An excess of water in the subgrade can also cause damage by freezing, causing frost heave in fine sands, chalk and silty soils. Conversely thawing may cause a reduction in subgrade strength giving rise to failure of the pavement above. Tree roots, particularly those of fast-growing deciduous trees such as the poplar and willow, can also cause damage in heavy clay soils by extracting vast quantities of water from the subgrade down to a depth of 3.000.

The pavement covering will give final protection to the subgrade from excess moisture but during the construction period the subgrade should be protected by a waterproof surfacing such as a sprayed bituminous binder with a sand cover applied at a rate of 1 litre per square metre. If the subgrade is not to be immediately covered with the sub-base of the pavement it should be protected by an impermeable membrane such as 500 gauge plastic sheeting with 300 mm side and end laps.

PAVEMENT CONSTRUCTION

Pavement is a general term for any paved surface and is also the term applied specifically to the whole construction of a road. Road pavements can be classified as flexible pavements which for the purpose of design are assumed to have no tensile strength and consist of a series of layers of materials to distribute the wheel loads to the subgrade. The alternative form is the rigid pavement of which, for the purpose of design, the tensile strength is taken into account and consists of a concrete slab resting on a granular base.

Flexible pavements

The sub-base for a flexible pavement is laid directly onto the formation level and should consist of a well-compacted granular material such as a quarry overburden or crushed rocks. The actual thickness of sub-base required is determined by the cumulative number of standard axles to be carried (where a standard axle is taken as 8 200 kg) and the CBR of the subgrade. The CBR, or California bearing ratio, is an empirical method in which the thickness of the sub-base is related to the strength of the sub-grade and to the amount of traffic the road is expected to carry. Data regarding CBR values can be obtained from the design charts contained in Road Note 29 published by HM Stationery Office.

The subgrade is covered with a sub-base, a base course and a wearing course; the last two components are called collectively the surfacing. The sub-base can consist of any material which remains stable in water such as crushed stone, dry lean concrete and blast furnace slag. Compacted dry-bound macadam in a 75 to 125 mm thick layer with a 25 mm thick overlay of firmer material or a compacted wet mix macadam in 75 to 150 mm thick layers could also be used. The material chosen should also be unaffected by frost and have a CBR value of not less than 80% and should be well compacted in layers giving a compacted thickness of between 100 and 150 mm for each layer.

The base course of the surfacing can consist of rolled asphalt, dense tarmacadam, dense bitumen macadam or open-textured macadam and should be applied to a minimum thickness of 60 mm. Base courses are laid to the required finished road section providing any necessary gradients or crossfalls ready to receive the thinner wearing course which should be laid within three days of completing the base course. The wearing course is usually laid by machine and provides the water protection for the base layers. It should also have non-skid properties, reasonable resistance to glare, have good riding properties and have a good life expectancy. Materials which give these properties include hot rolled asphalt, bitumen macadam, dense tar surfacing and cold asphalt — see Fig. VIII.2 for typical details. Existing flexible road surfaces can be renovated quickly and cheaply by the application of a hot tar or cut-back bitumen binder with a rolled layer of gravel, crushed stone or slag chippings applied immediately after the binder and before it sets.

The above treatments are termed bound surfaces but flexible roads or

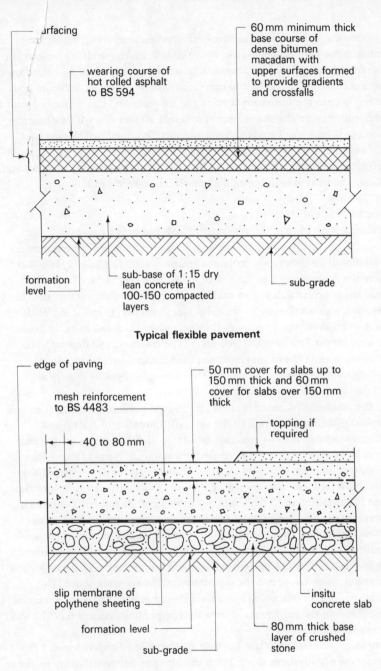

surfacing

wearing course of hot rolled asphalt to BS 594

60 mm minimum thick base course of dense bitumen macadam with upper surfaces formed to provide gradients and crossfalls

formation level

sub-base of 1 : 15 dry lean concrete in 100-150 compacted layers

sub-grade

Typical flexible pavement

edge of paving

mesh reinforcement to BS 4483

40 to 80 mm

50 mm cover for slabs up to 150 mm thick and 60 mm cover for slabs over 150 mm thick

topping if required

slip membrane of polythene sheeting

formation level

sub-grade

insitu concrete slab

80 mm thick base layer of crushed stone

Typical rigid pavement

Fig VIII.2 Typical pavement details

273

pavements with unbound surfaces can also be constructed. These are suitable for light vehicle traffic where violent braking and/or acceleration is not anticipated such as driveways to domestic properties. Unbound pavements consist of a 100 to 200 mm thick base of clinker or hardcore laid directly onto the formation level of the subgrade and this is covered with a well-rolled layer of screeded gravel to pass a 40 mm ring with sufficient sand to fill the small voids to form an overall consolidated thickness of 25 mm. This will give a relatively cheap flexible pavement but to be really efficient it should have adequate falls to prevent the ponding of water and should be treated each spring with an effective weed killer.

Rigid pavements

This is a form of road using a concrete slab laid over a base layer. The preparation of the subgrade is as described above for flexible pavements and should be adequately protected against water. The base layer is laid over the subgrade and is required to form a working surface from which to case the concrete slab and to enable work to proceed during wet and frosty weather without damage to the subgrade. Generally, granular materials such as crushed concrete, crushed stone, crushed slag and suitably graded gravels are used to form the base layer. The thickness will depend upon the nature and type of subgrade; weak subgrades normally require a minimum thickness of 150 mm whereas normal subgrades require a minimum thickness of only 80 mm.

The thickness of concrete slabs used in rigid pavement construction will depend upon the condition of the subgrade, intensity of traffic and whether the slab is to be reinforced. With a normal subgrade using a base layer of 80 mm thick the slab thickness would vary from 125 mm for a reinforced slab carrying light traffic to 200 mm for an unreinforced slab carrying a medium to heavy traffic intensity. The usual strength specification is 28 MN/m^2 at 28 days with not more than 1% test cube failure rate; therefore the mix design should be based on a mean strength of between 40 and 50 MN/m^2, depending upon the degree of quality control possible on or off site. To minimise the damage which can be caused by frost and de-icing salts the water/cement ratio should not exceed 0.5 by weight, and air entrainment to at least the top 50 mm of the concrete should be specified. The air entraining agent used should produce 3 to 6% of minute air bubbles in the hardened concrete thus preventing saturation of the slab by capillary action.

Before the concrete is laid the base layer should be covered with a slip membrane of polythene sheet which will also prevent grout loss from the concrete slab. Concrete slabs are usually laid between pressed steel road forms which are positioned and fixed to the ground with steel stakes.

These side forms are designed to provide the guide for hand tamping or to provide for a concrete train consisting of spreaders and compacting units. Curved or flexible road forms have no top or bottom flange and are secured to the ground with an increased number of steel stakes — see Fig. VIII.3 for typical road form examples.

Reinforcement generally in the form of a welded steel fabric complying with the recommendations of BS 4483 can be included in rigid pavement constructions to prevent the formation of cracks and to enable the number of expansion and contraction joints required to be reduced. If bar reinforcement is used instead of welded steel fabric it should consist of deformed bars at spacings not exceeding 150 mm. The cover of concrete over the reinforcement will depend on the thickness of concrete, for slabs under 150 mm thick the minimum cover should be 50 mm and for slabs over 150 mm thick the minimum cover should be 60 mm.

Joints used in rigid pavements may be either transverse or longitudinal and are included in the design to:

1. Limit the size of slab.
2. Limit the stresses due to subgrade restraint.
3. Make provision for slab movements such as expansion, contraction and warping.

The spacing of joints will be governed by a number of factors, namely slab thickness, presence of reinforcement, traffic intensity and the temperature at which the concrete is placed. Five types of joint are used in rigid road and pavement construction and are classed as follows:

1. *Expansion joints* — transverse joints at 36 to 72 m centres in reinforced slabs and at 27 to 54 m centres in unreinforced slabs.
2. *Contraction joints* — transverse joints placed between expansion joints at 12 to 24 m centres in reinforced slabs and at 4.5 to 7.5 m centres in unreinforced slabs to limit the size of slab bay or panel. It should be noted that every third joint should be an expansion joint.
3. *Longitudinal joints* — similar to contraction joints and are required where slab width exceeds 4.5 m.
4. *Construction joints* — the day's work should normally be terminated at an expansion or contraction joint, but if this is impossible a construction joint can be included. These joints are similar to contraction joints but with the two portions tied together with reinforcement. Construction joints should not be placed within 3.000 of another joint and should be avoided wherever possible.
5. *Warping joints* — transverse joints which are sometimes required in unreinforced slabs to relieve the stresses caused by vertical

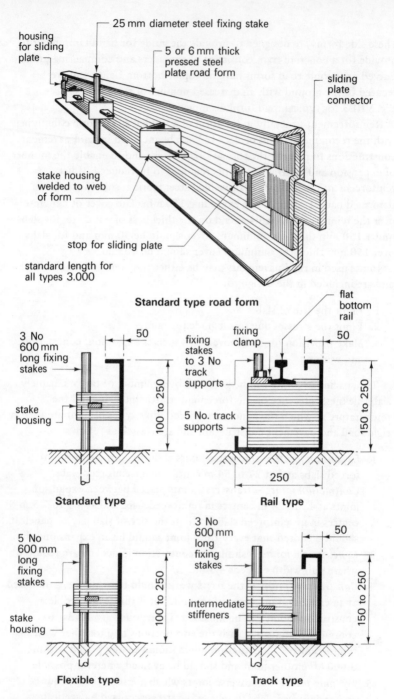

housing for sliding plate

25 mm diameter steel fixing stake

5 or 6 mm thick pressed steel plate road form

sliding plate connector

stake housing welded to web of form

stop for sliding plate

standard length for all types 3.000

Standard type road form

3 No 600 mm long fixing stakes

stake housing

50

100 to 250

Standard type

flat bottom rail

fixing stakes to 3 No track supports

fixing clamp

50

5 No. track supports

150 to 250

250

Rail type

5 No 600 mm long fixing stakes

stake housing

100 to 250

Flexible type

3 No 600 mm long fixing stakes

intermediate stiffeners

50

150 to 250

Track type

Fig VIII.3 Typical steel road form details

temperature gradients within the slab if they are higher than the contractional stresses. The detail is similar to contraction joints but it has a special arrangement of reinforcement.

Typical joint details are shown in Fig. VIII.4.

Road joints can require fillers and/or sealers; the former needs to be compressible whereas the latter should protect the joint against the entry of water and grit. Suitable materials for fillers are soft knot-free timber, impregnated fibreboard, chipboard, cork and cellular rubber. The common sealing compounds used are resinous compounds, rubber-bituminous compounds and straight run bitumen compounds containing fillers. The sealed surface groove used in contraction joints to predetermine the position of a crack can be formed whilst casting the slab or be sawn into the hardened concrete using water-cooled circular saws. Although slightly dearer than the formed joint, sawn joints require less labour and generally give a better finish.

The curing of newly laid rigid roads and pavings is important if the concrete strength is to be maintained and the formation of surface cracks is to be avoided. Curing precautions should commence as soon as practicable after laying, preferably within 15 minutes of completion by covering the newly laid surface with a suitable material to give protection from the rapid drying effects of the sun and wind. This form of covering will also prevent unsightly pitting of the surface due to rain. Light covering materials such as waterproof paper and plastic film can be laid directly onto the concrete surface, ensuring that they are adequately secured at the edges. Plastic film can give rise to a smooth surface if the concrete is wet; this can be avoided by placing raised bearers over the surface to support the covering. Heavier coverings such as tarpaulin sheets will need to be supported on frames of timber or light metal work so that the covering is completely clear of the concrete surface. Coverings should remain in place for about seven days in warm weather and for longer periods in cold weather.

DRAINAGE

Road drainage consists of directing the surface water to suitable collection points and conveying the collected water to a suitable outfall. The surface water is encouraged to flow off the paved area by crossfalls which must be designed with sufficient gradient to cope with the volume of water likely to be encountered during a heavy storm to prevent vehicles skidding or aquaplaning. A minimum crossfall of 1 : 40 is generally specified for urban roads and motorways whereas crossfalls of between 1 : 40 and 1 : 60 are common specifications for service roads. The run-off water is directed towards the edges of the

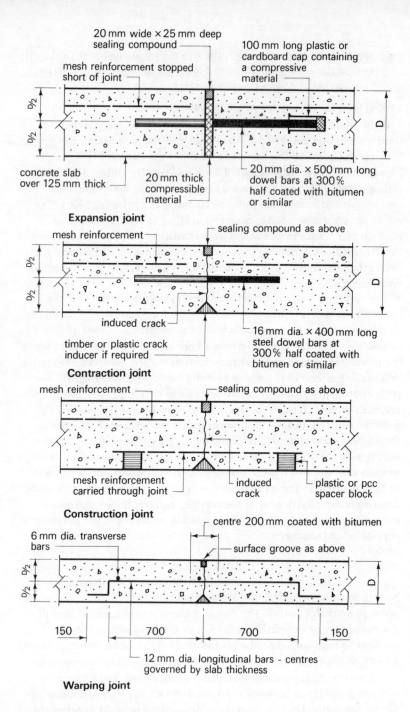

20 mm wide × 25 mm deep sealing compound

mesh reinforcement stopped short of joint

100 mm long plastic or cardboard cap containing a compressive material

concrete slab over 125 mm thick

20 mm thick compressible material

20 mm dia. × 500 mm long dowel bars at 300% half coated with bitumen or similar

Expansion joint

mesh reinforcement

sealing compound as above

induced crack

timber or plastic crack inducer if required

16 mm dia. × 400 mm long steel dowel bars at 300% half coated with bitumen or similar

Contraction joint

mesh reinforcement

sealing compound as above

mesh reinforcement carried through joint

induced crack

plastic or pcc spacer block

Construction joint

6 mm dia. transverse bars

centre 200 mm coated with bitumen

surface groove as above

150   700   700   150

12 mm dia. longitudinal bars - centres governed by slab thickness

Warping joint

Fig VIII.4 Typical road joint details

road where it is in turn conveyed by gutters or drainage channels at a fall of about 1 : 200 in the longitudinal direction to discharge into road gullies and thence into the surface water drains.

Road gullies are available in clayware and precast concrete with or without a trapped outlet; if final discharge is to a combined sewer the trapped outlet is required and in some areas the local authority will insist upon a trapped outlet gully for all situations. Spacing of road gullies is dependent upon the anticipated storm conditions and crossfalls but common spacings are 25 to 30 m. The gratings are usually made from cast iron, slotted and hinged to allow easy flow into the collection chamber of the gully and to allow for access for suction cleaning — see Fig. VIII.5. Roads which are not bounded by kerbs can be drained by having subsoil drains beneath the verge or drained directly into a ditch or stream running alongside the road. The sizing and layout design of a road drainage system is normally covered in the services syllabus of a typical building course of study. It is for this reason such calculations are not included in this text.

FOOTPATHS AND PEDESTRIAN AREAS

These can be constructed from a wide variety of materials or in the case of a large area they can consist of a mixture of materials to form attractive layouts. Widths of footpaths will be determined by local authority planning requirements but a width of 1.200 is usually considered to be the minimum in all cases. Roads are usually separated from the adjacent footpath by a kerb of precast concrete or natural stone which by being set at a higher level than the road not only marks the boundary of both road and footpath but also acts as a means of controlling the movement of surface water by directing it along the gutter into the road gullies — see Fig. VIII.6.

Flexible footpaths, like roads, are those in which for design purposes no tensile strength is taken into account. They are usually constructed in at least two layers consisting of an upper wearing course laid over a base course of tarmacadam. Wearing courses of tarmacadam and cold asphalt are used, the latter being more expensive but having better durability properties — see Fig. VIII.6 for typical details. Gravel paths similar in construction to unbound road surfaces are an alternative method to the layered tarmacadam footpaths. Loose cobble areas can make an attractive edging to a footpath as an alternative to the traditional grass verge. The 30 to 125 mm diameter cobbles are laid directly onto a hardcore or similar bed and are handpacked to the required depth.

Unit pavings are a common form of footpath construction consisting of a 50 to 75 mm thick base of well-compacted hardcore laid to a minimum

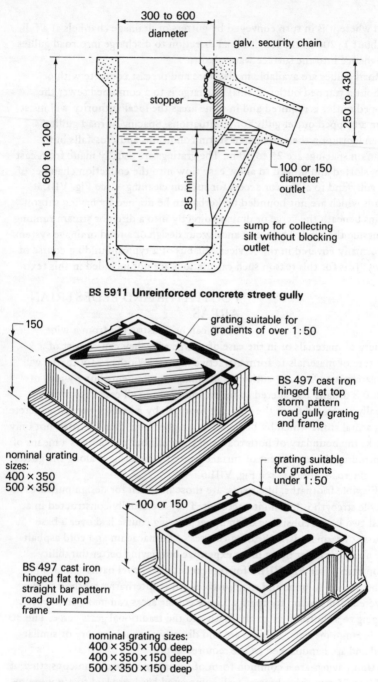

300 to 600
diameter

galv. security chain

stopper

250 to 430

600 to 1200

85 min.

100 or 150
diameter
outlet

sump for collecting
silt without blocking
outlet

BS 5911 Unreinforced concrete street gully

150

grating suitable for
gradients of over 1:50

BS 497 cast iron
hinged flat top
storm pattern
road gully grating
and frame

nominal grating
sizes:
400 × 350
500 × 350

grating suitable
for gradients
under 1:50

100 or 150

BS 497 cast iron
hinged flat top
straight bar pattern
road gully and
frame

nominal grating sizes:
400 × 350 × 100 deep
400 × 350 × 150 deep
500 × 350 × 150 deep

Fig VIII.5 Typical road gully and grating details

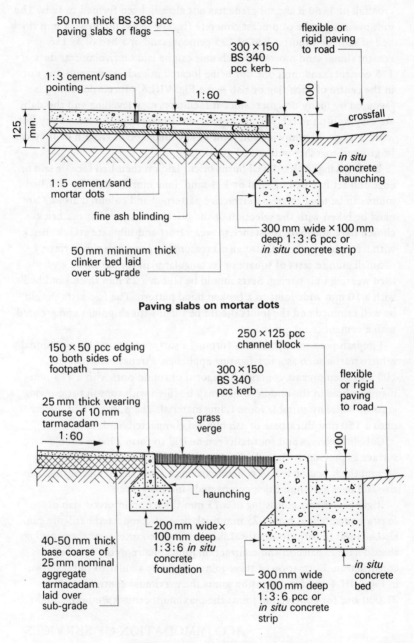

Paving slabs on mortar dots

50 mm thick BS 368 pcc paving slabs or flags

1:3 cement/sand pointing

1:60

300 × 150 BS 340 pcc kerb

flexible or rigid paving to road

100

crossfall

125 min.

in situ concrete haunching

1:5 cement/sand mortar dots

fine ash blinding

300 mm wide × 100 mm deep 1:3:6 pcc or *in situ* concrete strip

50 mm minimum thick clinker bed laid over sub-grade

Flexible footpath with grass verge

150 × 50 pcc edging to both sides of footpath

250 × 125 pcc channel block

300 × 150 BS 340 pcc kerb

flexible or rigid paving to road

25 mm thick wearing course of 10 mm tarmacadam

1:60

grass verge

100

haunching

200 mm wide × 100 mm deep 1:3:6 *in situ* concrete foundation

40-50 mm thick base coarse of 25 mm nominal aggregate tarmacadam laid over sub-grade

300 mm wide ×100 mm deep 1:3:6 pcc or *in situ* concrete strip

in situ concrete bed

Fig VIII.6 Typical footpath details

281

crossfall of 1 : 60 if the subgrade has not already been formed to falls. The unit pavings can be of precast concrete flags or slabs laid on a 25 mm thick bed of sand or a mortar bed of 1 : 5 cement : sand or a bed of 1 : 1 : 6 cement : lime : sand mortar or each unit can be laid on five mortar dots of 1 : 5 cement : sand mix, one dot being located in each corner and one dot in the centre of each flag or slab − see Fig. VIII.6. Mortar dot fixing is favoured by many designers since it facilitates easy levelling and the slabs are easy to lift and relay if the need arises. Paving flags can have a dry butt joint, a 12 to 20 mm wide soil joint to encourage plant growth or they can be grouted together with a 1 : 3 grout mix.

Brick pavings of hard well-burnt bricks laid on their bed face or laid on edge and set in a bed of sand or 1 : 4 sand : lime mortar with dry or filled joints can be used to create attractive patterned and coloured areas. Care must be taken with the selection of the bricks to ensure that the bricks chosen have adequate resistance to wear, frost and sulphate attack. Bricks with a rough texture will also give a reasonably good non-slip surface.

Small granite setts of square or rectangular plan format make a very hard wearing unit paving. Setts should be laid in a 25 mm thick sand bed with a 10 mm wide joint to a broken bond pattern. The laid setts should be well rammed and the joints should be filled with chippings and grouted with a cement : sand grout.

Firepath pots are suitable for forming a surface required for occasional vehicle traffic such as a fire-fighting appliance. Firepaths consist of 100 mm deep precast concrete hexagonal or round pots with a 175 mm diameter hole in the middle which may be filled with topsoil for growing grass or with any suitable loose filling material. The pots are laid directly onto a 150 mm thick base of ash blinded compacted hardcore.

Cobbled pavings and footpaths can be laid to form a loose cobble surface as previously described or the oval cobbles can be hand set into a 50 mm thick bed of 1 : 2 : 4 concrete having a small maximum aggregate size laid over a 50 mm thick compacted sand base layer.

Rigid pavements consisting of a 75 mm thick unreinforced slab of *in situ* concrete laid over a 75 mm thick base of compacted hardcore can also be used to form footpaths. Like their road counterparts these pavings should have expansion and contraction joints incorporated into the construction. Formation of these joints would be similar to those shown in Fig. VIII.4 and for expansion joints the maximum centres would be 27.000 and for contraction joints the maximum centres would be 3.000.

ACCOMMODATION OF SERVICES

The services which may have to be accommodated under a paved area could include any or all of the following:

1. Public or private sewers.
2. Electrical supply cables.
3. Gas mains.
4. Water mains.
5. Telephone cables.
6. Television relay cables.
7. District heating mains.

In planning the layout of these services it is essential that there is adequate co-ordination between the various undertakings and bodies concerned if a logical and economical plan and installation programme is to be formulated.

Sewers are not generally grouped with other services and due to their lower flexibility are given priority of position. They can be laid under the carriageway or under the footpath or verge, the latter position creating less disturbance should repairs be necessary. The specification as to the need for ducts, covers and access positions for any particular service will be determined by the undertaking or board concerned. Most services are laid under the footpath or verge so that repairs will cause the minimum of disturbance and to take advantage of the fact that the reinstatement of a footpath is usually cheaper and easier than that of the carriageway.

Services which can be grouped together are very often laid in a common trench commencing with the laying of the lowest service and backfilling until the next service depth is reached and then repeating the procedure until all the required services have been laid. The selected granular backfilling materials should be placed in 200 to 250 mm well-compacted layers. All services should be kept at least 1.500 clear of tree trunks, any small tree roots should be cut, square trimmed and tarred. Typical common service trench details are shown in Fig. VIII.7.

INTERNAL SLABS AND PAVINGS

The design and construction of a large paved area to act as a ground floor slab is similar to that already described in the context of rigid pavings. The investigation and preparation of the subgrade is similar in all aspects to the preparatory works for roads. A sub-base of well-compacted graded granular material to a thickness of 150 mm for weak subgrades and 80 mm thick for normal subgrades should be laid over the formation level. If the sub-base is to be laid before the roof covering has been completed the above thicknesses should be increased by 75 mm to counteract the effects of any adverse weather conditions. A slip membrane of 500 gauge plastic sheeting or similar material should be laid over the sub-base as described for rigid pavings for roads.

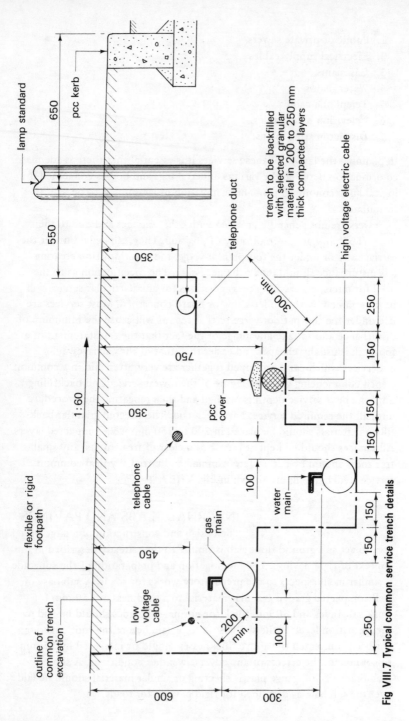

Fig VIII.7 Typical common service trench details

The *in situ* concrete slab thickness is related to the load intensity, required life of the floor and the classification of the subgrade. Floors with light loadings of up to 5 kN/m² would require a minimum thickness of 175 mm for weak subgrades and a minimum thickness of 150 mm for normal subgrades, whereas heavier slab loadings would require minimum slab thicknesses of 200 mm for weak subgrades and 175 mm for normal subgrades. The grade of concrete specified would range from 20 N/mm² for offices and shops to 40 N/mm² for heavy industrial usages. The need, design, detail and spacing of the various types of joints is as described for rigid pavings and detailed in Fig. VIII.4.

Large internal pavings or slabs are laid in bays to control the tensile stresses due to the thermal movement and contraction of the slab. Laying large areas of paving in bays is also a convenient method of dividing the work into practicable sizes. Bay widths of 4.500 will enable a two-man tamping beam to be used, and will facilitate the easy placing and finishing of the slab. This is also an ideal width for standard sheet of welded steel fabric reinforcement. Two basic bay layouts are possible, namely chequered board and long strip. Chequered board layout is where alternate bays in all directions are cast with joints formed to their perimeters. The intermediate bays are laid some seven days later to allow for some shrinkage to take place. Although this is the traditional method it has two major disadvantages in that access to the intermediate or infill bays is poor and control over shrinkage is suspect.

The long strip method of internal paving construction provides an excellent alternative and is based on established rigid road-paving techniques. The 4.500 wide strips are cast alternately which gives good access; expansion joints if required are formed along the edges; transverse contraction joints can be formed as the work proceeds or the surface groove can be sawn in at a later time. With this method it may be necessary to cast narrow edge strips some 600 to 1.000 wide at the commencement of the operation if perimeter columns or walls would hinder the use of a tamping board or beam across the full width of the first strip.

The concrete paving must be fully compacted during laying and this can be achieved by using a single or double tamping beam fitted with a suitable vibration unit. The finish required to the upper surface will depend upon the usage of the building and/or any applied finishes which are to be laid. Common methods employed are hand trowelling using steel trowels, power floating and power trowelling. The latter methods will achieve a similar result to hand trowelling but in a much shorter space of time. None of the above finishing methods can be commenced until the concrete has cured sufficiently to accept the method being used.

Vacuum dewatering is a method of reducing the time delay before power floating can take place. The object is to remove the excess water from the slab immediately after the initial compaction and levelling has taken place. The slab is covered with a fine filter sheet and a rigid or flexible suction mat to which is connected a transparent flexible plastic pipe attached to a vacuum generator. The vacuum created will compress the concrete slab and force the water to flow out up to a depth of 300 mm. The dewatering process will cause a reduction of about 2% in the slab depth and therefore a surcharge should be provided by means of packing strips on the side forms or at the ends of the tamping boards. The filter sheet will ensure that very little of the cement fines of the mix are carried along in suspension by the water being removed. The vacuum should be applied for about three minutes for every 25 mm of concrete depth which will generally mean that within approximately 20 minutes of casting the mats can be removed and the initial power floating operation commenced and followed by the final power trowelling operation. This method enables the laying of long strips of paving in a continuous operation. It is possible for a team consisting of six operatives and a ganger to complete 200 m^2 of paving per day by using this method.

Bibliography

Relevant BS — British Standards Institution.

Relevant BSCP — British Standards Institution.

Building Regulations 1985 — HMSO.

Relevant BRE Digests — HMSO.

Relevant Advisory Leaflets — DOE.

DOE Construction Issues 1—17 — HMSO.

R. Barry. *The Construction of Buildings*. Crosby Lockwood and Sons Ltd.

Mitchells Building Construction Series. B. T. Batsford Ltd.

W. B. McKay. *Building Construction*, Vols. 1 to 4. Longman.

W. Fisher Cassie and J. H. Napper. *Structure in Buildings*. The Architectural Press.

Cecil C. Handisyde. *Building Materials*. The Architectural Press.

Handbook on Structural Steelwork — The British Constructional Steel Work Association Ltd and the Constructional Steel Research and Development Organisation.

B. Boughton. *Reinforced Concrete Detailers Manual*. Crosby Lockwood and Sons Ltd.

G. N. Smith. *Elements of Soil Mechanics for Civil and Mining Engineers*. Crosby Lockwood and Sons Ltd.

Construction Safety. The National Federation of Building Trades Employers.

A Guide to Scaffolding Construction and Use. Scaffolding (Great Britain) Ltd.

B.S.P. Pocket Book. The British Steel Piling Co Ltd.

R. Holmes. *Introduction to Civil Engineering Construction.* College of Estate Management.

Lighting for Building Sites. The Electricity Council.

Relevant A J Handbooks. The Architectural Press.

Relevant manufacturers' catalogues contained in the Barbour Index and Building Products Index Libraries.

Index

290

291